AMERICA IN THE GREAT WAR

AMERICA IN THE GREAT WAR

The Rise of the War Welfare State

RONALD SCHAFFER

New York   Oxford
OXFORD UNIVERSITY PRESS
1991

Oxford University Press

Oxford   New York   Toronto
Delhi   Bombay   Calcutta   Madras   Karachi
Petaling Jaya   Singapore   Hong Kong   Tokyo
Nairobi   Dar es Salaam   Cape Town
Melbourne   Auckland

and associated companies in
Berlin   Ibadan

Library of Congress Cataloging-in-Publication Data
Schaffer, Ronald.
America in the Great War : the rise of the war welfare state
Ronald Schaffer.
p. cm. Includes bibliographical references and index.
ISBN 0-19-504903-9
1. United States—Politics and government—1913–1921.
2. World War, 1914–1918—United States.
I. Title
E780.S34 1991 973.91'3—dc20 91-8316

9 8 7 6 5 4 3 2 1

Printed in the United States of America
on acid-free paper

To my parents

Preface

This book is about the way the United States government led the American nation through nineteen months of the First World War. It focuses on two themes: the government's efforts to manage the American people, encouraging and enabling them to support and fight a massive war in Europe; and the use Americans made of that war. It is about a nation struggling for ideals rich with meaning—democracy, personal freedom, and lasting peace. It also tells of individuals and organizations trying to make the Great War serve their particular, sometimes highly materialistic purposes.

Certain topics usually found in books on war—the causes of the conflict, details of tactics and weapons, diplomatic maneuvering, and the peace settlement—are not the concerns of this work and are discussed only enough to provide a framework for the central arguments; yet these pages contain much information about the impact of war on a nation and its inhabitants that will prove, I believe, as fascinating to the reader as it has to me.

Most of this information derives from the published literature discussed in the Essay on Sources. Some of it, particularly in the sections on propaganda and on combat, is based on unpublished materials. While I have written *America in the Great War* primarily for general readers of history and for undergraduate students, an audience that might find detailed notes superfluous, anyone who wants to know the origins of specific points can borrow a fully cited copy from the California State University, Northridge library.

I would like to acknowledge the financial support that California State University, Northridge has given to this project and the assistance of several individuals: librarians and archivists at CSUN, the University of Southern California, the Hoover Institution, the New York Public Library, and the University of California, Los Angeles; Richard J. Sommers, archi-vist-historian at the U.S. Army Military History Institute; John Broesamle, Ronald Davis, Paul Koistinen and Leonard Pitt, who read various drafts of the manuscript; Gloria Richards, who helped enter data from the World War I survey; Regina Lark-Miller and Marcia Dunnicliffe of the CSUN history department; Sheldon Meyer and Scott Lenz at Oxford University Press; and students in my classes on war and society who evaluated several chapters. The responsibility for any errors and all interpretations is mine.

January 1991   R. S.
Northridge, Calif.

Contents

Introduction

Toward the end of June 1914, a story appeared in American newspapers describing the murder of a European prince, heir to the throne of the Austro-Hungarian Empire. This event, so remote to the people of the United States, precipitated a general European conflict that quickly spread to other parts of the world and eventually drew in their own nation. The Great War, as contemporaries called it, was the first of the man-made catastrophes of the twentieth century, a conflict that killed some twenty million people, wounded millions more, and left untold numbers ruined. It ignited revolutions in eastern and central Europe and greatly changed the societies of those belligerents that did not collapse, including the United States.

The social effects of the Great War stemmed, in part, from the way it was fought. It was a conflict of entire peoples in which millions of soldiers, backed by the civilians of their nations, battled both for military objectives and to bleed the enemy into surrendering—the kind of war in which victory depended not just on the courage, weapons, and skill of those who fought but also on the productivity and endurance of people at home. The belligerent governments had to manage their nations in such a way as to make troops persist, year after year, slaughtering one another in the front lines, while inducing noncombatants to feed the conflict with material and emotional sustenance.

The United States government faced an extremely demanding task in mobilizing its own nation for this war. Decades later, with their country mired in such limited conflicts as those in Korea and Vietnam, some Americans longed for an age when their nation fought total wars like World War I, supposedly without restraints. But even in that earlier conflict, when the American commander in chief implored his nation to apply "Force, Force, to the utmost, Force without stint or limit," all kinds of hindrances restricted America's ability to fight.

The government responded to these limits with exhortation, by offering rewards for service, and with threats and coercion. Drawing on deep
national reservoirs of patriotism, it used a massive propaganda campaign to
stimulate public enthusiasm and to arouse hostility against the nation's
foreign and domestic enemies. It tapped into a popular tradition of voluntary action, offering psychic and material rewards to citizens for willing
service and directing great pressure against those who held back. The
government prosecuted and imprisoned individuals who questioned the
war or tried to impede its activities; yet it ran the wartime nation more by
manipulation than by direct control, through incentives more than
punishments.

Although the wartime government disseminated its benefits widely, it
gave the largest rewards to the powerful, to the rich and the organized, to
those whose services and products it most urgently needed, and to those
most skillful in obtaining what it had to offer. A system of bargaining
developed in which the government tried to manage interest groups while
the interest groups used the war to extract benefits for themselves or for
those they represented. In a conflict waged ostensibly for ideals like freedom and democracy, patriotism and idealism flourished alongside an intense pursuit of personal and corporate gain and of social objectives only
indirectly related to the war. The result was a war welfare state, a phenomenon that became, as the historian Lawrence Stone observed, the
norm for advanced societies during the twentieth century.

Ordinarily, people think of a welfare state as one in which government
takes responsibility for matters affecting poor and middle-class citizens,
providing social security, health care, housing and education, and subsis-
tence payments for the poor and setting minimum-wage levels for working
people. But in the United States, wealthy individuals and corporations
have received government benefits as much as, or perhaps more than,
other elements of society. The war welfare state of 1917-1918 was rooted
in a tradition of public generosity. Since the beginning of the Republic,
American governments had provided rewards and protections to citizens
and businesses—in such forms as tariffs, grants of land and loans, internal
improvements, and removing the native population from the path of white
settlement. It was the scope of government aid that sets off the war years
from earlier times.

We will examine here, facet by facet, the system that evolved in wartime for managing American society. After introducing the chief administrator, President Wilson, and explaining briefly how the United States
became a party to the conflict, we will describe his administration's program for making a public that for the most part had wanted to stay out of
the conflict mentally ready to sustain a great war. We will discuss the

program's impact on the American people, including the way it fostered a frenzy of intolerance, and then analyze the struggle to mobilize the American economy, emphasizing the incentives government provided to induce industry to convert to production for war. A series of chapters describes the effects of the war on reform movements, showing how government and reformers used and aided one another and why some of those movements flourished in the war years and others did not. Intellectuals played important parts in wartime America. We will discuss them at some length, explaining what motivated those who supported the war, how the Wilson administration aided and exploited them, and what happened to certain intellectuals who dissented. In another set of chapters, we will explore the government's efforts to raise and inspire an American army, letting the letters, diaries, and memoirs of soldiers speak for those men about the conditions they had to face in France, and we will show how their government and its agents encouraged them to enter the battlefield and fight effectively and tried to help those with wounded minds go back into combat. A brief epilogue connects wartime themes to the later development of American society.

The War President

Woodrow Wilson, the man who led the United States in the First World War, was born in 1856 in Staunton, Virginia, into a deeply religious family of Scottish and Scotch-Irish extraction, people without much wealth but with high standing in their community. His father, a Presbyterian minister, was a sociable man with a gift for fluent self-expression. Wilson's mother, who had descended from a family of learned scholars and Calvinist divines, was proud and strong and so reserved that, as her son recalled, "only those of her own household can have known how lovable she was." Young Thomas Woodrow (as he was then called) was tutored by his father and entered Davidson College at seventeen. After a year, he dropped out and spent several months at home, reading and discussing great men with a friend in his hometown. Then he entered Princeton University and after graduation studied law at the University of Virginia Law School; however, the "haggling and scheming" practice of that profession, as he described it, did not suit him, and in 1883 he entered the Johns Hopkins University to study history and political science.

After graduation, Wilson began a rapid ascent, publishing a series of well-received books, teaching at Bryn Mawr, Wesleyan University, and finally at Princeton. Students found him a brilliant, inspiring lecturer. In 1902 he was named president of Princeton and quickly established a reputation as one of the country's leading educators, raising standards and changing the way undergraduates were taught. His efforts to integrate a new graduate school into the university and to abolish the Princeton eating clubs (social organizations to which many upperclassmen belonged) ended in failure, and he resigned in 1910 to begin a career in politics. He was elected governor of New Jersey that year and in 1913 began the first of two terms as president of the United States.

Woodrow Wilson was a man of pronounced contradictions. His controlled posture and facial expression, the buttoned-up suits he wore, and the carefully contrived rhythms of his speeches often made him seem the essence of formality. Yet beneath his grave surface, a different kind of person resided, observable in the ironic smile that could sometimes be seen in the mask of his long bony face and the vitality that sparkled through his blue-grey eyes, in the emotional letters he wrote to loved ones, in the humor he displayed in private, and in the passion of his speeches.

Like an actor, he spoke and behaved and probably felt the way his part required. Once he warned a group of listeners not to confuse him with the presidency. Observing that the office he occupied seemed much greater than himself, he remarked that all he could do was "to look grave enough and self-possessed enough to seem to fill it." But he added that he could "hardly refrain, every now and again, from tipping the public a wink, as much to say, 'It is only *me* that is inside this thing.'" The "thing" was not just the presidency; it was also the façade that Wilson presented to the world.

In his political thinking and behavior, one observes pronounced conflicts and contradictions. An aura of idealism enveloped him, and his advisors understood that when they wanted to bring up some practical policy, it was wise to frame it in principled arguments; yet he operated in a highly pragmatic way, shifting positions when it was expedient to do so, deceiving his listeners at times, acting, in other words, like a normal politician. Thus, at a meeting during his campaign for governor of New Jersey, with the leader of the Democratic machine that was helping elect him sitting in the audience, he solemnly declared, "[I]f you find out that I have been or ever intend to be connected with a machine of any kind I hope you will vote against me." Stiff-necked and domineering, particularly under stress, so stubborn at times that scholars have wondered if his rigidity resulted from a neurological disorder, Wilson could also negotiate and compromise. The closest advisors and appointees of this autocratic president included such persons as Colonel Edward M. House and Bernard M. Baruch, whose methods were subtle, indirect, and manipulative.

Wilson became the world's most celebrated champion of democracy. In private, he was a snob, bored by the ordinary citizens of his country. He told his fiancée during his first term as president that the great majority of people who came to his office, the majority or even the minority of congressmen, and most American voters were "not of our kind." In a history he wrote of the American people, he disparaged immigrants from southern and eastern Europe—men of "the lowest class" and "of the meaner sort," men "out of the ranks where there was neither skill nor energy nor any initiative of quick intelligence." He distrusted people who came to the United States from China and looked down on Americans of African origin. Yet millions of ordinary people throughout the world responded to the idealism he preached and thought of him as their advocate.

From a foe of unruly labor unions, centralized government, and agrarian radicals and socialists, Wilson metamorphosed into a reformer, announcing, as he sought progressive votes in his race for governor, "I am and always have been an insurgent." He became the opponent of inefficient trusts; an advocate of banking reform; a supporter of credits for farmers, the vote for women, and opportunities for independent businessmen; and an enemy of child labor. During a struggle with some of Princeton's wealthy trustees, in an outburst of emotion, he told a reporter that directors of corporations and trusts were like "burglars legally organized for plunder." Yet he never seriously challenged basic economic arrangements in the United States, and as president he appointed to federal regulatory agencies persons who sympathized with the business community.

Wilson's foreign policies reflected his appreciation of American private enterprise. "Lift your eyes to the horizons of business," he told a salesmanship congress.

> [L]et your thoughts and your imaginations run abroad throughout the whole world, and with the inspiration of the thought that you are Americans and are meant to carry liberty and justice and the principles of humanity wherever you go, go out and sell goods that will make the world more comfortable and more happy, and convert them to the principles of America.

The great project of his second administration was to create an international system that would allow American capitalism and America's kind of political freedom to spread peacefully across the globe.

In his dealings with other nations, Wilson was a realist, which, in the United States of his time, with so many idealistic inhabitants, meant justifying policies in legalistic and moralistic terms. It meant proclaiming that "America was born into the world to do mankind service" and that Americans had made themselves "the champions of free government and national sovereignty." Realism also meant starting a massive naval building program while the country was at peace and beginning preparations for an

expanded army, and it included a willingness to use force.° Wilson em-
barked on more military interventions than any other American president
before or since, seven including World War I. In the Western Hemi-
sphere, pursuing American economic and strategic interests, he sent
armed forces into Haiti and Cuba—countries that lay athwart a sea route
to the Panama Canal—into Haiti's neighbor, the Dominican Republic,
and into Mexico, where German agents were fomenting hostility against
the United States.

After the outbreak of war in Europe, Wilson's administration proclaimed
American neutrality but began taking actions that made it, before 1917, an
undeclared participant. Although both the Allies and the Central Powers
used blockades against their enemies, the United States government allowed its citizens to travel into war zones on ships that sometimes carried
munitions, insisting that international law and traditions gave it the right to
do so. It also permitted American manufacturers to ship materials that
helped maintain the economies of warring nations and allowed American
lenders to extend credit to belligerents, actions that helped bring the
United States out of a recession and were promoted within the Wilson
administration as a way of fostering a war boom. Since the Allies, through
control of the ocean surface, took almost exclusive advantage of American
trade and financing, the United States became their arsenal and banker.
The German response, which at that time seemed extraordinarily inhumane to many Americans, was to employ submarines against ships that
sustained their enemies, torpedoing them without warning; thus they destroyed American property and took American lives.

After a crisis brought on by the sinking of the British liner *Lusitania*
and by other U-boat attacks, the administration was able to pressure the
German government into restricting its submarines to surface warfare, a
method often fatal to these slow-moving, fragile warships. Wilson, who
wished to see neither side achieve an absolute victory over the other, tried
to arrange a settlement before direct American military intervention became necessary; but the belligerents did not intend to yield at a peace
conference what they had won with blood, and they hoped that additional
killing would secure their objectives. Late in 1916 the Germans calculated
that they could shut off American supplies to the British and cripple the
Allies with renewed submarine attacks before America's armed forces
could make a difference in the war. After a series of sinkings, the stifling of
American trade, and a revelation of German intrigues against the United
States in Mexico, Wilson asked Congress to declare war against Germany,

°This is not meant to imply that Wilson did not believe in the ideals he proclaimed.

asserting that his country would fight unselfishly, without rancor, for democracy and self-government, for the liberties of small nations, and for a league of nations that would bring peace and freedom to the entire world. On April 6, 1917, the United States became an avowed belligerent. The Wilson administration's task was now to ready the American people, emotionally and physically, to take part in what their president called "the most terrible and disastrous of all wars."

AMERICA IN THE GREAT WAR

1 ‖ Managing American Minds

Almost from the beginning of hostilities in Europe, President Wilson had imagined how difficult it would be to lead a divided country into the Great War. He recognized that in such a conflict unity was vital and internal opposition, or even apathy, might be ruinous. As the United States took its place among the belligerents, he initiated programs to shape the way Americans perceived the war, to provide them information the government wished them to know, and to motivate them. Meanwhile, his administration set out to silence dissenters.

The president had good reasons to worry about the way the public felt. He had been elected the previous fall with the slogan, "He Kept Us Out of War." Now, although millions of Americans were rushing to enlist in the armed forces or volunteer for civilian service, others accepted the war without enthusiasm, did not understand the reasons for fighting, or preferred not to take part. There were signs of disaffection among farmers in the South and Midwest and among Scandinavian and Russian emigrants who had left the Old World to escape from war or to avoid military service. Irish-Americans were torn between their ties to the United States and their hatred of England. Many German- and Austrian-Americans not only felt the pull of ancestry but also feared they might have to face their own relatives on the battlefield. The influence of pacifists, some of whom had been demonstrating in the streets of Washington almost to the moment war was declared, was incalculable, and many progressives, even strong supporters of Wilson, feared the war would unravel domestic reforms. Americans on the radical left, meanwhile, considered the Great War an imperialist struggle, a last upheaval before the disintegration of capitalism.

For months after war was declared, indications of disunity and apathy continued to reach the White House. At the end of May, the president's secretary, Joseph P. Tumulty, told him how discouraging it was to find the general mass of people "indifferent." Secretary of State Robert Lansing added that many, perhaps the great majority, of the American people required an emotional stimulus "to keep them up to the proper pitch of earnestness and determination. . . ." When officials of a federal war agency wrote prominent Americans in the summer of 1917 calling their attention to a lack of feeling for war, Roy Howard, the head of a major newspaper chain, replied: "This weakness must be remedied before the nation will go to war with its heart as well as its hands and feet." Secretary of War Baker remarked at a cabinet meeting in August that the people did not know what they were fighting for and urged the president to tell them.

Wilson could not wait for enthusiasm to grow spontaneously. Soon after the U.S. entry into the war, he learned that Great Britain and France, on the verge of losing their Russian ally and menaced by a tightened U-boat blockade and the threat of German land attacks in western Europe, needed immediate help. If the United States were to sustain the Allies, avert an enemy breakthrough, and uphold its influence among the warring nations, the administration would have to generate mass support for mobilization. It needed some mechanism for shaping the way Americans thought and felt about the Great War.

George Creel and the Committee on Public Information

On April 13, the president issued an executive order establishing a Committee on Public Information (CPI) to conduct American propaganda in the United States and overseas. He named as its directors the secretaries of war, state, and the navy, provided it with several million dollars from executive funds, and appointed dynamic forty-one-year-old publicist and social reformer George Creel to manage it.

Creel had grown up in the American heartland, in Independence and Odessa, Missouri. He was a progressive journalist, a writer of "uplift" poetry, the co-author of an exposé of child labor (*Children in Bondage*), and a publicity man for the New York City Woman Suffrage party. With a group of other reformers, he had set about to clean up the city of Denver, where, as police commissioner, he had shut down its red-light districts. In the prewar years, Creel, like many progressives, had wanted to take power away from an alliance of corrupt politicians and corporations and promoted direct-democracy measures like the referendum to restore it to ordinary citizens. An early enthusiastic booster of Woodrow Wilson, Creel was eager to run the campaign to promote the war, even though he believed the other powers were fighting for trade and empire and felt the strongest impetus to involve the United States had come from the most reactionary and least democratic elements in American society. He wanted to help Wilson transform the World War into a struggle for democracy.

Creel's committee tried to manage American thinking about the war without the kind of direct, sweeping censorship that other belligerent nations imposed. It based its methods on techniques previously devised by politicians, commercial advertisers, and Allied propaganda experts. Superficially simple and direct, these methods were actually quite sophisticated.

In his account of the CPI's activities, "the world's greatest adventure in advertising," as he called them, Creel painted a picture of a government leveling with its people, simply and straightforwardly giving them facts. His version is partly true. The Wilson administration did supply the American people with facts. By doing so, it reassured them that they were being kept informed and helped prevent wild and dangerous rumors. Its policy of presenting facts was also intended to avoid the kind of embarrassment that occurs when forbidden information leaks out.

Yet the Creel committee did not assume that the public would simply take the facts it disseminated and draw the right conclusions. It regarded public opinion as at least partly irrational and tried to direct it with appeals to the emotions. At one point, Creel remarked that his aim was to weld Americans into "one white-hot mass . . . with fraternity, devotion, courage, and deathless determination." This was not likely to occur if people subjected the facts they received to reasoned evaluation.

The way the government handed out information made rational analysis difficult. The CPI and other agencies overwhelmed the public with details. By the end of 1917, they were sending each newspaper in California an average of six pounds of publicity a day. It was impossible for the newspapers or their readers either to verify this flood of details or to analyze the information and draw independent conclusions. The people who provided the publicity supplied the conclusions, frequently in a simplified way designed to touch the emotions.

While the CPI usually identified itself as the source of its publicity, it sometimes preferred to keep its audience unaware that it was trying to persuade them. The head of the Creel committee's film division remarked that one of the committee's problems was to disseminate "telling propaganda which at the same time would not be obvious propaganda, but will have the effect we desire to create." The committee obscured its purposes by employing euphemisms (for example, calling the government's propaganda agency the "Committee on Public Information") and by excluding the word "propaganda" from its public statements. Although the CPI played a crucial role in censorship, it avoided the terms "censorship" or "censor" when communicating with outsiders, and its chairman hoped that the flood of information the CPI supplied would keep people from noticing that censorship was taking place. The Creel committee also tried to create the impression that the materials it sent out originated from a source congenial to the recipients. For this purpose, it established front organizations—the Friends of German Democracy as a conduit for its publicity to German-Americans; and the American Alliance for Labor and

Democracy, nominally headed by the union leader Samuel Gompers, to influence American workers.

The committee released its torrent of facts through all kinds of media. It issued press releases, placed ads in newspapers and magazines, organized conferences and ran speaking campaigns, attached posters to walls and billboards all over America, manufactured cartoons and slides, and produced war movies. With CPI assistance, private motion-picture producers created films with such provocative titles as *The Bath of Bullets, The American Indian Gets into the War Game,* and *A Girl's a Man for A' That* (about women war workers). With cooperating organizations, the CPI staged dramatic newsworthy events: exhibitions, pageants, and parades. To arouse the enthusiasm of civilians, it used warplanes and thousands of troops to stage mock battles (when men and planes were desperately needed at the front).

The CPI tried to communicate with every segment of American society. To recent immigrants, it passed out leaflets in their native languages. Its Division of Women's War Work placed information in newspapers and magazines about the efforts of women to win the war. To reach those who could not or would not read, the committee sent thousands of Four-Minute Men to motion-picture theaters, where they delivered patriotic messages or led the audience in patriotic songs during the four minutes it took to change reels.

Creel's organization worked closely with educators. College professors in the Division of Civic and Educational Cooperation wrote propaganda pamphlets, a *War Cyclopedia,* and other publications to educate the American people about why their country was fighting. (We will look into their operations in a later chapter.) They prepared essay contests for high schools and an outline for study of the war. The educational division also printed little war stories and poems for elementary school children and encouraged teachers to tell their pupils tales that illustrated patriotism, heroism, and sacrifice—stories about children of invaded countries, about horses, Red Cross dogs, and carrier pigeons.

Themes of American Propaganda

In the great variety of messages the government disseminated, three major themes appear: a unity motif, the image of a despicable enemy, and the idea of a crusade for peace and freedom. National unity was a crucial matter to the president, who tried to strengthen it with words. "I pray God," Wilson declared a few weeks after the country had entered the war, "that some day historians will remember these momentous years as the years which made a single people of the great body of those who call themselves Americans." In May 1917 he observed that Americans were in a war of entire nations and urged them to act as a "team." And later, when public feeling was more soundly behind the war, he declared that a hundred years of peace could not have knitted the country together as a single year of war had done.

To potential sources of disaffection, the government spread its exhortations for national solidarity. The CPI distributed German-language pamphlets on loyalty to the United States and passed out over half a million copies of a leaflet, "Friendly Words to the Foreign Born," in German, Bohemian, Italian, Russian, Hungarian, and English. Government artists painted the idea of loyalty with such posters as "America Gave You All You Have to Give," "Remember Your First Thrill of American Liberty," and "Remember! The Flag of Liberty. Support It!" all depicting immigrant types about to debark in the golden land. The Creel committee's National School Service news sheet encouraged teachers to develop in their students, whether of native or foreign birth, a "unity of thought" that would break down barriers between children of recent and those of earlier immigrant parentage.

The federal government directed special messages to farmers and workers. Creel's committee distributed an appeal from the president to the farmers of America and presidential speeches about the need for wartime harmony between labor and capital. Wilson tried to counter the belief, thought to be widespread among workers, that capitalists who stayed home were reaping unconscionable profits while ordinary working men went off to fight in France. The nation, Wilson declared, was fighting a laborers' war. It was "taking counsel with no special class. . . . serving no private or single interest." To reinforce this claim, the CPI produced a leaflet called "Who Is Paying for this War" and had labor historian John R. Commons draw up "Why Workingmen Support the War," which explained that the profits of capitalists were being controlled and that union leaders served on the boards that set wages and hours. The American Alliance for Labor and Democracy published more than three hundred thousand copies of Professor Common's pamphlet.

Government motion pictures portrayed farmers and laborers cooperating in the war effort. In a tableau from one film, we see standing side by side a workingman and a soldier who together are holding a toy ship; an aviator supporting a two-bladed propeller; a sailor, stripped to the waist, who carries a shell cartridge; and a farmer, scythe in hand, standing behind a large lumpy sack that appears to contain some kind of vegetable. In the foreground, a nurse ministers to a wounded soldier while Columbia, a formidable-looking woman clad in a long, nightgown-like garment with a steel helmet on her head, stands at the rear, holding an American flag with one hand and pointing with the other over the heads of the figures in front. This scene was part of *America's Answer,* a nine-reel epic filmed by the Army Signal Corps and distributed by the CPI to more than a third of the nation's motion-picture theaters.

When it tried to convey an image of the enemy itself, the administration was inconsistent. At first, it sought to distinguish an innocent German people from a guilty German government, an approach that would help shield German-Americans from the wrath of their neighbors and align them with the United States. In practice, the distinction broke down. A CPI pamphlet, "Conquest and Kultur: Aims of the Germans in Their Own Words," implied that Germans in general were guilty of their country's transgressions. "One may not draw an indictment against a whole nation," the pamphlet observed. But by quoting statements from what it described as responsible political and intellectual leaders of Germany who had "shaped the thought and action of the German people in the past generation," "Conquest and Kultur" blurred the difference between German citizens and their rulers.

America's propaganda image of Germany reflected German military actions in the European war—attacks on civilians in parts of Belgium, destruction of cultural centers (particularly the university town of Louvain), and U-boat sinkings of ocean liners—as magnified by Allied propaganda and by the imagination of American artists and writers. The image varied in emotional intensity. It ranged from Wilson's phrases about "ruthless, lawless" German submarine warfare to hideous atrocity stories spread by the Crcel committee in millions of pamphlets and advertisements. One can imagine the feelings of New York University graduates as they read an ad, "Bachelor of Atrocities," that the CPI inserted in their alumni magazine:

> In the vicious guttural language of Kultur, the degree A.B. means Bachelor of Atrocities. . . . The Hohenzollern° fang strikes at every element of decency and culture and taste that your college stands for. It leaves a track so terrible that only whispered fragments may be recounted. It has ripped all the world-old romance out of war, and reduced it to the dead black depths of muck, and hate, and bitterness. . . . Are you going to let the Prussian Python strike at your Alma Mater? . . .

This advertisement, although directed at the highly educated few, contained one of the CPI's frequently used images—of the enemy as animal-like, or at least devoid of human qualities. This image usually

---

°The Hohenzollerns were the ruling family of the Kingdom of Prussia and after 1871 of the German Empire.

combined beastliness with sex or violence or both. Thus in the poster "Halt the Hun!" a stern American doughboy restrains with his forearm a subhuman German soldier whose hand rests on the shoulder of a ragged, kneeling, comely woman who has a naked infant in her lap. "This Is Kultur" shows a German soldier holding a boy by the neck with one hand and grasping the stump of the boy's wrist with another, a second soldier holding the bloody sword that severed the child's hand, and in the background a third German strangling a young woman. (The secretary of the CPI advertising division thought this poster should be placed in the *Ladies' Home Journal.*)

Some of the more lurid posters added fantasies about enemy attacks on the United States. In "Destroy this Mad Brute," a monstrous wild-eyed ape with a Kaiser Wilhelm mustache and huge drops of saliva dripping from its gaping mouth stands on a piece of land labeled "AMERICA." It carries with its left arm (which has matted fur and blood on it) a light-skinned woman whose long grey dress has been torn down, leaving her breasts exposed. The woman covers her face with one hand while holding the side of her head with the other. The ape wields a club labeled "KULTUR" and its spiked helmet is inscribed with the word "MILITARISM." "The Hohenzollern Dream" depicts a gigantic, portly German soldier, also with a Kaiser Wilhelm mustache, holding a rifle and bayonet on one shoulder while a vulture perches on the other. His left boot is partly immersed in New York harbor, and with his right boot he is crushing the towers of skyscrapers. Joseph Pennell's poster "That Liberty Shall Not Perish from the Earth" shows an attack by two- and four-winged aircraft that has left lower Manhattan in flames. A submarine lurks in waters nearby, and in the foreground stands a wrecked Statue of Liberty, whose severed head rests partly submerged.

To avenge such atrocities and to secure lasting peace and bring democracy to other nations, American propaganda urged the public to join a crusade against German militarist autocracy. President Wilson called for

> Force, Force to the utmost, Force without stint or limit, the righteous and triumphant Force which shall make Right the law of the world, and cast every selfish dominion down in the dust.

On the first draft-registration day, he told an assembly of Confederate veterans how this crusade was part of God's plan. People in other countries, he said, had thought of the United States as a "trading and money-getting people"; but in reality, God had made the United States indestructible and indivisible, not so the American people could achieve their own selfish objectives but so they could carry out His work. "And now," he said,

" we are to be an instrument in the hands of God to see that liberty is made secure for mankind." Wilson accused the Central Powers of outraging the principles of "knightly honor." Of front-line American troops he said, "They are crusaders . . . fighting for no selfish advantage for their own Nation," and proclaimed that Americans were combating German militarism in the cause of peace.

Secretary of State Lansing explained why the crusade was a war for peace:

> Were every people on earth able to express their will, there would be no wars of aggression, and, if there were no wars of aggression, then there would be no wars, and lasting peace would come to this earth. The only way that a people can express their will is through democratic institutions. Therefore, when the world is made safe for democracy . . . universal peace will be an accomplished fact.

American propaganda echoed these views of Wilson and Lansing. The Great War was a "crusade," the Creel committee declared, "not merely to re-win the tomb of Christ, but to bring back to earth the rule of right, the peace, goodwill to men and gentleness he taught." An advertising poster for *Pershing's Crusaders,* a government motion picture, shows General John J. Pershing, commander of the American Expeditionary Force, erect on horseback and followed by a color guard and by rank on rank of doughboys.* Ghostly medieval knights ride in the background with red crosses on their shields. On another poster, a squad of doughboys prepares to bayonet fleeing German troops while Columbia stands nearby, one hand pointing at the Germans and the other at Christ, crucified on a cross above the Americans.

Perhaps the best summation of the crusade theme appears in the last stanzas of "The New Crusade" (by Katherine Lee Bates), which the CPI distributed in a collection of patriotic prose and poetry:

> Sons of the granite,
>   Strong be our stroke,
> Making this planet
>   Safe for the folk.
>
> Life is but passion,
>   Sunshine on dew.
> Forward to fashion
>   The old world anew!

---

*A name given to American troops.

"A Nation of Traders"!
   We'll show what we are
Freedom's crusaders
   Who war against war.

Images of Women in American Propaganda

In the imagery of war propaganda, America's crusading knights offered their lives to protect women, like the young, long-haired girl being dragged by a spike-helmeted, mustached German soldier across the Liberty Loan poster "Remember Belgium," or the mother saved from the ape-like enemy in "Halt the Hun!" But the imagined women also played quite different roles. They were Dame Liberty and Dame Victory in the *Stars and Stripes,* the newspaper for American soldiers, which told its readers "to hold all women as sacred." They appeared as heroines like the smiling Joan of Arc encased in armor with a look of inspiration in her eyes, holding a sword before her in a war stamp poster "Joan of Arc Saved France. Women of America Save Your Country." They were shown as stern goddesses—as Columbia in *America's Answer* and in recruiting posters like "The Navy Needs You! Don't READ American History. MAKE IT"; as the Statue of Liberty, who glares and points at the viewer from C. R. Macauley's poster "You Buy a Liberty Bond Lest We Perish"; and as "Public Opinion," dressed in a filmy garment, a cape and sandals, and a cap labeled "PUBLIC OPINION," who confronts the viewer, one fist clenched before her, saying,

> I Am Public Opinion
> All men fear me!
> I declare that Uncle Sam shall not go to his knees to beg you to buy his bonds. That is no position for a fighting man. But if you have the money to buy and do not buy, I will make this No Man's Land for you!

War artists depicted women as healthy, self-sufficient gardeners, as determined war workers, as elderly ladies pleading with American women to buy Liberty Bonds, or as vivacious, sexy "girls." A woman Red Cross nurse appeared in farm and trade papers and women's magazines as "The Greatest Mother in the World," cradling in her arms a tiny wounded soldier, the size of a baby, who lies with his head bandaged on a tiny stretcher, protected by a blanket.

These are a few examples from the campaign to sell the Great War to the American people. To gauge in a rough way the magnitude of that campaign, one must take into account the incidence of government-staged events (like Liberty Bond parades) and the enormous number of reproductions of each bit of propaganda ("The Greatest Mother in the World," for instance, appeared more then fifty million times). One must also imagine how often Americans saw or heard the government's messages. When they looked at a billboard, opened a newspaper, or examined a magazine, government propaganda was likely to greet them. At the movies, the chances were good that they would see a CPI film, for in a country with about 12,000 theaters, *Pershing's Crusaders* was booked into nearly 4,100, *America's Answer* was shown in nearly 4,500, and an "Official War Review" in newsreel format appeared 6,950 times. Even at ordinary commercial movies, there were the Four-Minute Men rousing war spirit between reels. Walking on Main Street, Americans would see one of the CPI's sixty thousand patriotic window displays or perhaps encounter a Liberty Bond rally or some other occurrence intended to shape their thinking about the war. When the doorbell rang, it might be a boy scout delivering a copy of the president's latest speech. Speaking to acquaintances or listening to a sermon in church, one might hear a government message filtered through another mind.

These efforts at informing and persuading the people of the United States comprised only one side of the government's efforts to mobilize American thought and feeling about the war. There was another side, a suppressive and punitive side, to the management of American minds.

2 ||| Controlling Dissent

Almost from the start of hostilities in Europe, President Wilson had been concerned about what would happen to American liberties if the United States ever entered the Great War. He knew from his study of history how patriots had persecuted loyalists in the American Revolution, about the "odious" Sedition Act of 1798 that cut, he wrote, "perilously close to the root of freedom of speech and of the press," and of Lincoln's harsh measures against dissenters in the Civil War. He believed that because of the way human nature operated, war would inevitably lead to "moral exhaustion" and a loosening of the principles of just and temperate government. His own wartime administration amply justified these concerns. As temporary measures for exceptional circumstances, it controlled what the American people could read in newspapers and magazines and see in motion pictures, silenced dissidents, and encouraged public officials and private persons to regulate the way Americans thought and behaved.

Its methods were simultaneously manipulative and authoritarian. Rejecting proposals to have the armed forces run a censorship program, the administration tried to induce "voluntary" self-censorship and sought, meanwhile, to create so much enthusiasm for the war that people would not notice or resent the way information was being suppressed. But against those who did not cooperate with its voluntary program, it employed sweeping coercive powers promulgated through executive orders or granted by Congress in the Espionage Act of June 1917, the Trading with the Enemy Act of October 1917, and the Sedition Act of May 1918.*

Shortly after the United States entered the war, the Committee on Public Information suggested to editors that if they had any doubts about an article they intended to print, they should submit it for approval. Since the CPI did not tell the editors specifically what was banned, they had to guess what the government might object to; and unless they were unusually courageous or naïve, they interpreted the committee's directive broadly, because the committee also explained that if they published any-

---

*See Appendix for the relevant sections of these laws.

thing it did not like, it would turn them in to the Post Office and the Department of Justice, agencies with specific authority to curb dissent.

Sometimes the CPI itself intervened to decide what Americans could see or read. George Creel asked the Librarian of Congress to remove the antiwar book *War—What For?* from the shelves and to send him the names of people who had asked for it. At Creel's behest, a Censorship Board (on which he served) prepared lists of suspect magazine and book publishers, whose works were banned from export. The CPI also limited what could be shown in motion pictures. Newsreel companies that wanted to shoot at any military installation had to apply to the committee for a permit, allow it and military intelligence agencies to review the film, and secure a release before exhibiting what resulted. The CPI warned motion-picture companies that it would keep films it disapproved of out of lucrative overseas markets. Because the Essanay Film Company's *The Curse of Iku* seemed to evoke prejudice against the people of Japan, a friendly country, the committee, together with the State and War departments forced the producers to re-edit it before allowing it to be shown.

In the Espionage Act, Congress authorized the Post Office to deny the mails to any piece of writing that violated the act's other provisions. The Post Office used these powers vigorously. The day after the act was signed, its director, Albert S. Burleson, asked all local postmasters to send him any unsealed mail, "newspapers, etc." that might violate the act or hamper or "embarrass" the government in its conduct of the war. The Trading-with-the-Enemy Act gave Burleson's agency authority to make foreign-language newspapers supply local postmasters with English translations of any comments on a wide variety of subjects relating to the war. Since translation was time-consuming and expensive, the burden on those papers was substantial, forcing some of the smaller ones to close and others to print innocuous information. But it was applied selectively—mainly against left-wing newspapers like the socialist *Philadelphia Tageblatt*.

By the middle of July 1917, Burleson and Post Office solicitor William H. Lamar had excluded from the mails issues of the *Masses*, the *International Socialist Review*, the *Appeal to Reason*, the *American Socialist*, the *Milwaukee Leader*, and other publications. While some of these journals were openly critical of the war, Burleson also suppressed an issue of the *Public* for urging that more money be raised by taxes and less by loans, and he censored the *Freeman's Journal and Catholic Register* for reprinting Jefferson's opinion that Ireland should be a republic. On the ground that a periodical that missed one issue no longer qualified for second-class mailing privileges, he denied those privileges to the *Masses*.

People asked Burleson what his standards were for censoring publications. At first he refused to reply, but eventually he announced that a publication could not be loyal if it said "that this Government got in the war wrong, that it is in for the wrong purposes, or anything that will impugn the motives of the Government for going into the war"; nor could it say that the federal government was the tool of Wall Street or the munitions makers or anything that would "obstruct the Government in the prosecution of the war." These ideas, Burleson said, made for insubordination in the armed services and bred a spirit of disloyalty in the country. It was the view of the postmaster general that when the majority had spoken, legally and properly, "in its wisdom and patriotism . . . every loyal member of the minority should become one with the majority."

While the Post Office was attempting to weed out literature, federal law-enforcement officials moved against suspected opponents of the war. The Justice Department instituted nearly 2,200 prosecutions and secured 1,055 convictions. Among its accomplishments was the conviction of more than ninety-six leaders of the Industrial Workers of the World (or "Wobblies"), a militant, radical labor organization that not only interfered with production of a vital war material by striking copper mines but also openly promulgated antiwar ideas. Judge Kenesaw Mountain Landis sentenced the IWW official William D. Haywood to twenty years in prison and a $30,000 fine. The Socialist party leader Eugene V. Debs, who had received over nine hundred thousand votes in 1912 as candidate for president, was given a ten-year sentence for a vaguely antiwar speech in which he urged his listeners to join his party and said, "You need to know that you are fit for something better than slavery and cannon fodder."

Most of those prosecuted were much less radical than Debs and Haywood but not sufficiently discreet. J. P. Doe, son of the chief justice of New Hampshire, was imprisoned for writing a chain letter in which he denied that Germany had broken the *Sussex* Pledge, a 1916 agreement by the German government not to engage in unrestricted submarine attacks provided that the United States compelled the British navy to adhere to international law. Since the British navy had followed its own version of international law, Doe's argument was highly plausible; but that did not help him. He received an eighteen-month sentence for trying to impede recruiting. Moviemaker Robert Goldstein was convicted of attempting to cause insubordination in the armed forces. Goldstein had produced *The Spirit of '76*, an epic film of the Revolutionary War. Along with Patrick Henry's "Give Me Liberty" speech and the signing of the Declaration of Independence, he had included a scene of the notorious Wyoming Massacre, showing British and Hessian soldiers bayoneting women and children and dragging women by the hair, and an English officer carrying a young woman into his bedchamber. A federal judge gave him ten years, commuted in 1919 to three. The film was re-edited into a patriotic feature under court supervision. Goldstein's company went bankrupt, losing over $100,000.

D. H. Wallace, a former British soldier, received twenty years for saying that when a soldier went away he was a hero but when he came back he was a bum and that the asylums would be filled with veterans. Wallace had also incriminated himself by stating that the troops were giving their lives for the capitalists and that 40 percent of Allied guns and ammunition was defective because of graft. Wallace went insane and died in prison. A judge gave Ohio farmer John White twenty-one months for saying that soldiers in American camps were dying like flies and that the murder of innocent women and children by German soldiers was no worse than what American troops had done to native people during the U.S. occupation of the Philippines. Walter Matthey of Iowa received a one-year term for attending a meeting at which someone else attacked conscription.

A brewer, a banker, and a shoemaker in Covington, Kentucky, were sentenced to five, seven, and ten years respectively after private detectives, using a hidden dictaphone, recorded conversations in which they allegedly criticized the Red Cross, expressed pleasure at German military success, and called Theodore Roosevelt a "damned agitator." A married man with five children and little money who had been forced to buy a Liberty Bond was turned in by a relative for saying, during an argument, that "this is a rich man's war," that atrocity stories were lies, and that he hoped the government "goes to hell. . . ." The judge fined him $5,000 and sentenced him to twenty years, later reduced to two. A Texan went to the penitentiary for saying, "Wilson is a wooden-headed son of a bitch. I wish Wilson was in hell, and if I had the power I would put him there."

One of the most publicized cases involved socialist Rose Pastor Stokes. Mrs. Stokes had written the Kansas City *Star,* denying that she believed every citizen should give unqualified support to the United States in its war aims. "No government which is for the profiteers," she wrote, "can also be for the people, and I am for the people while the Government is for the profiteers." Judge Arba S. Van Valkenburgh sentenced her to ten years for words that, in his view, might cause insubordination or mutiny and tended to lessen enthusiasm for the war among relatives of servicemen. Valkenburgh remarked that the success of America's armed forces depended on the support and aid of "the folks at home."

A small number of federal judges, including Learned Hand and Charles F. Amidon, construed the Espionage Act narrowly, allowing the accused some right to speak their minds. In Amidon's North Dakota district court, during one term, a grand jury returned only thirteen valid indictments in over a hundred fifty cases; but in most of the wartime free-speech cases, judges, prosecutors, and juries acted with less restraint.

Justice Oliver Wendell Holmes, Jr., who was extremely reluctant himself to rule against the government in these cases, later characterized the lower federal courts during this period as "hysterical." In the Haywood case, Judge Landis remarked that the peacetime right to speak against preparing for war or going to war ceased to exist once war was declared. The United States attorney for the Southern District of Ohio observed that the courts were "not permitting persons who have taken a stand against the Government in this war to seek shelter behind the Constitution" and that in the wartime crisis the Constitution had been "virtually suspended." In the Stokes case, Judge Van Valkenburgh defined freedom of speech as the protection of "criticism which is . . . friendly to the government, friendly to the war, friendly to the policies of the government."

To help it track down opponents of war, the Justice Department employed a quarter of a million volunteers organized in the American Protective League. Carrying "Secret Service Division" cards, APL members impersonated federal officers, opened private mail, wiretapped telephones, recorded conversations with dictaphones, broke into offices, conducted searches and seizures, and delivered materials they uncovered to the Justice Department's Bureau of Investigation. Together with bureau agents, APL'ers raided German-language newspapers. They investigated aliens and socialists, reported disloyal remarks, and entrapped suspects. From time to time they called private citizens into their offices and lectured them on how to be loyal. For the War Department, the Red Cross, the Knights of Columbus, and the YMCA, they conducted tens of thousands of character and loyalty investigations. As a result of these inquiries, which were not subject to any formal appeal, some eight hundred persons were barred on grounds of disloyalty from taking Civil Service examinations.

In its most spectacular operation (the New York City "slacker raids" of September 3–6, 1918), members of the league working with Department of Justice agents hauled suspected draft dodgers out of office buildings, lofts, and street cars and detained them for judgment by draft boards. The raids netted 50,187 persons, of whom 16,505 were judged to have violated selective-service rules. The American Protective League did much to justify the claim of Assistant Attorney General John Lord O'Brian that the United States had never been more thoroughly policed.

While federal agents and their auxiliaries moved against offenders, state and local authorities, often in cooperation with the federal government, sought to stifle antiwar, radical, and pro-German attitudes. Nine states made it a crime to express opposition to the war effort. Several state governments cooperated with the American Protective League or local vigilante groups to track down and punish people suspected of disloyalty.

In most regions, the investigating and enforcement agency was a council of defense (sometimes called a public-safety committee or loyalty bureau) staffed by volunteers who received authority from state or local governments. By 1918 there were 184,000 of these organizations arranged in a country-wide hierarchy, with a federal Council of National Defense at its apex.* The councils performed numerous valuable war services. They helped the federal government raise money in Liberty Bond sales, took part in programs to produce more food and consume less of it, organized teams of women to knit sweaters for American servicemen, and encouraged able-bodied unemployed men to go to work or join the armed forces. State and local councils also promoted draft registration, assisted searches for army deserters, and helped the federal government fight prostitution around army camps.

Councils of defense roused patriotic feeling and became conduits for administration propaganda. In Connecticut, for example, the state organization dispatched hundreds of speakers to conduct war rallies and address meetings throughout the state. It sent information to presumed makers of public sentiment—chiefly clergy, executives of industrial corporations, bankers, and influential aliens. One product of the Connecticut council, a poster entitled the "Prussian Blot," showed a map of the German Empire spreading from Berlin to Baghdad. The council's publicity department sent "Liberty Choruses" into foreign-speaking communities to sing patriotic songs in factories, churches, schools, and meeting halls.

State legislatures delegated extraordinary powers to these volunteer organizations. The Minnesota Public Safety Commission received authority to compel people to appear before it with their records and testify under oath. It could even investigate any state official and, if it so decided, recommend dismissal. Montana lawmakers provided their council of defense with authority to do everything necessary and proper for public safety and to protect life and property—not just public property, but any private property the council chose to protect. The Montana legislature also declared that the state council of defense, unless specifically prohibited by the Montana or United States constitutions, could do anything else required to prosecute the war. Acting under this immense grant of authority, the Montana council asserted the right to hold hearings and investigations and the power to subpoena testimony and compel production of evidence. These powers could be used to reward or punish. The Montana state organization and its county branches received permission to pass out over half a million dollars to farmers from a state bond issue. At the same time,

*For the other activities of the Council of National Defense, see chapter 3.

anyone who violated the council's orders faced a fine of up to a thousand dollars, a year in jail, or both.

Throughout America, councils of defense fought dissent. The Indiana Council called in suspects for interviews. Nebraska's council of defense appointed a Secret Service Department, which included every sheriff in the state, to ferret out disloyal inhabitants. Local committees heard their cases and referred those who failed to cooperate to the state council for further action. Suspects were not permitted counsel or told who their accusers were. A few were ordered to report at prescribed intervals to the state attorney general. The Oklahoma Loyalty Bureau sought out "systematically disloyal people," tried to educate those who did not know better, and jailed people who refused to conform. In California, the state council asked each of its county branches to gather information on disloyal individuals. As a result of these investigations, some two hundred people were interned in 1917 on charges of disloyalty or sedition.

At the suggestion of the state council, New Mexico county councils of defense printed form letters to tell various people that they had made remarks unbecoming to persons "enjoying the liberty and protection of the United States." These letters declared that such conduct would not be tolerated and warned recipients that they were "under surveillance." The Missouri state council printed three kinds of warning cards for those who were reported to have acted disloyally: a white card for first-time offenders, a blue card for the second offense, and a red card ordering the person who received it to report to the local postmaster.

Councils of defense required people to contribute money to the war. In South Dakota, the state organization advised county councils to subpoena persons who appeared to be buying less than their quota of Liberty Bonds and interrogate them about their financial resources. Residents of Iowa who failed to purchase the assigned amount were summoned before "slacker courts." The Minnesota Public Safety Commission appointed investigators for each county to determine if individuals had made "proper subscriptions." The names of those who failed to subscribe to wartime fund drives were placed on special note cards. In Michigan, merchants who did not buy enough bonds were told they would be labeled "slacker" or "pro-German."

Americans were encouraged to spy on one another and report evidence of disloyalty and pacifism. A children's Anti-Yellow Dog League listened for antiwar comments. In Washington, a state-sponsored group called the Minute Men planted agents in schools to report on instructors who taught the German language or "Hun" history courses. It received reports from a University of Washington professor concerning his colleagues on the faculty. With over twelve thousand members in Seattle alone, the Minute Men conducted more than two thousand investigations. A Kansas council-of-defense handbook urged its readers to report immediately any person or family who criticized the government in a negative way, questioned the government's purpose, or talked against the war. The Tulsa, Oklahoma, *Daily World* told readers: "Watch your neighbor. If he is not doing everything in his power to help the nation in this crisis, see that he is reported to the authorities." The Helena, Montana, *Independent* announced on its front page, "YOUR NEIGHBOR, YOUR MAID, YOUR LAWYER, YOUR WAITER MAY BE A GERMAN SPY." The *Literary Digest* asked readers to send it editorials they felt were seditious or treasonable.

The chief targets of all these activities were more than eight million German-Americans, the country's second largest national minority. Many German-Americans were very proud of their cultural heritage, and during the war their opponents tried to expunge the slightest traces of it. Americans ceased to eat frankfurters or sauerkraut or keep dachshunds as pets. They now consumed "liberty sausages," "liberty cabbage," and walked "liberty dogs." Both Ludwig van Beethoven and Fritz Kreisler were banned in Pittsburgh, and for a while the Philadelphia Orchestra and the Metropolitan Opera stopped playing German music altogether. Karl Muck, the conductor of the Boston Symphony Orchestra, who had agreed only reluctantly to begin concerts with the "Star-Spangled Banner," was arrested by federal agents as he was about to conduct Bach's "The Passion According to Saint Matthew" and interned as an undesirable alien.

By the summer of 1918, almost half the states had restricted or banned the use of the German language. The South Dakota council of defense ordered the state's elementary and secondary schools to stop teaching it and prohibited people from using it in telephone conversations. There were numerous burnings of German-language books, at least nineteen in Ohio. German-speaking church members were ordered to conduct their services in English. In Iowa, the governor issued a proclamation forbidding conversation in any language but English in public places, on trains, and over telephones and told those who could not speak or understand English to conduct religious worship in their homes. "The official language of the United States and the state of Iowa," he declared, "is the English language. Freedom of speech is guaranteed by federal and state constitutions," but not the right to speak in any other language than English.

In Montana, the state council of defense led the campaign against German culture, banning the German language from public and private schools and telling schools and libraries to ban textbooks containing German propaganda. The council listed twelve specific books that Montanans were forbidden to read, one of them Willis M. West's *Ancient World,* which council members considered too favorable toward the old Germanic tribes. Some school authorities cheerfully complied. The principal of Brockway High School told the council that he had taken the liberty of discontinuing the study of German and had the students burn their German books. In May 1918, Mrs. Emil Peterson of the town of Hilger wrote the Montana council:

> Last year we "weeded" out all german texts that were in our school library, clipped out all german songs in our book of national songs, blotted out the coat of arms and german flags in the dictionaries, and urged that every home should destroy german text and library books they possess. We also spell germany without a capital letter. A few days ago we burned all our West's ancient Worlds, and I have the permission of our trustees to destroy any texts found to contain german propaganda.

In the spring of 1918, as German armies launched a great offensive on the Western Front and casualty reports from American units reached the United States, hostility against German-Americans and opponents of war exploded. Terrible Threateners, Sedition Slammers, Knights of Liberty, and other named and nameless patriotic organizations terrorized their enemies. Near the end of March, John H. Wintherbotham, the midwestern field representative of the Council of National Defense state-councils section reported:

> All over this part of the country men are being tarred and feathered and some are being lynched. . . . These cases do not get into the newspapers nor is an effort ever made to punish the individuals concerned. In fact, as a rule, it has the complete backing of public opinion. . .

These are some of the incidents that occurred during the three months after Wintherbotham made his report:

> ▶ On the main street of Kenney, Illinois, several men beat William Heiseman, a farmer, for allegedly making pro-German remarks. Two American flags were nailed to the outside of his home, and he was told that if he removed them he would be tarred and feathered.

> ▶ A band of young men forced Morris Gotler, a merchant of East Alton, Illinois, to kiss the American flag and threatened him with hanging. He had ignored a merchants' agreement to close their doors during a demonstration promoting the sale of Liberty Bonds.

▶ In Corpus Christi, Texas, a German Lutheran pastor was whipped for preaching in German in defiance of an order from the county council of defense.

▶ In Oakland, California, members of the Knights of Liberty hanged a German-American tailor for a moment; but before he strangled, they cut him down and tied him to a tree.

▶ Reverend A. F. Meyer, pastor of a German Lutheran congregation near Pinckneyville, Illinois, entered a St. Louis hospital for treatment of injuries. Five men had attacked him and coated him with tar for allegedly unpatriotic remarks he had made in a sermon.

▶ In Athens, Illinois, a "loyalist committee" forced John Rynders, a grocer, to swear allegiance to and kiss a flag and then tied it around his neck. Rynders had refused to contribute groceries for a community bond-drive dinner.

▶ A Pensacola, Florida, citizens' vigilance committee flogged a German-American, forced him to shout "To hell with the kaiser; hurrah for Wilson," and ordered him to leave the state.

▶ Four drunken inhabitants of Pocahontas, Illinois, aimed a revolver at John Paulaites and forced him to kiss a flag they had pulled off the wall of his home.

▶ George Koetzer was tarred, feathered, and chained to a cannon in a San Jose, California, park for alleged pro-German behavior.

▶ An Austrian-American in Avoca, Pennsylvania, accused of criticizing the Red Cross, was tied up, hoisted thirty feet in the air, and sprayed with a fire hose. After an hour he was released.

▶ In San Rafael, California, a man accused of disloyalty had his hair cut into the shape of a cross and then was tied to a tree on the courthouse lawn.

▶ Several hundred men and boys ducked Dr. J. C. Biemann of La Salle, Illinois, into a canal. Later they forced him to kiss a flag. Biemann was thought to have called Secretary of War Baker a fathead.

▶ Students at Rutgers University stripped an antiwar Socialist student who had refused to speak at a Liberty Bond rally, covered him with molasses and feathers, and forced him to parade blindfolded through the streets. Signs carried at the head of the procession read, "This is what we do with pro-

Germans!" "He's a Bolsheviki!" and "He is against the Liberty Loan and the U.S.A."

One incident that attracted national attention was the murder of Robert P. Prager, a German-American drifter. When the United States declared war, Prager took out his first citizenship papers and although blind in one eye tried to enlist in the navy. The navy turned him down. After registering for the draft in St. Louis, he moved to the small town of Collinsville, Illinois, where he worked as a baker until his stubborn, argumentative manner got him fired. He found a temporary job as a miner in Maryville, a nearby community. On April 3, 1918, a group of miners seized him, denounced him as a spy, and paraded him through town. When they released him, he went home to Collinsville but returned to Maryville the next day to post a document in which he appealed for his rights and proclaimed his loyalty to the United States.

That night, a mob of about seventy-five Maryville miners left a saloon and went to Prager's house. "Brothers," he said to them, "I am a loyal U.S.A. workingman." They dragged him into the street, pulled off his shoes and his outer clothing, draped a flag on him, and made him march in front of them singing a patriotic song. At this point, the police intervened. An officer ran with Prager to the police station, but mob members followed and got into the building, where they found him hiding in the basement in a pile of sewer tiles. They forced him to parade again and made him sing patriotic songs and kiss the flag. Then, nine members of the mob took him out of town. Someone brought a towing rope. One end was placed around a tree; the other, around Prager's neck. At this point, Prager received permission to write a letter to his parents in Dresden, Germany, in which he told them he was about to die and asked for their prayers. Then he said to his killers: "All right boys, go ahead and kill me, but wrap me in the flag when you bury me." At 12:30 A.M. he was yanked ten feet into the air and died.

Opponents of German culture struck at German-speaking religious denominations, particularly Mennonites, who were doubly vulnerable because they felt that to buy Liberty Bonds, serve in the armed forces, or support war in any way violated God's commands. Pastor R. E. Vomhof of Laurel, Montana, where German was banned in churches, asked the state council of defense for permission to administer communion in the language of his Russian-German congregation because it was "sinful to partake in the Lord's supper without understanding."* The council refused

*Russian Germans were Germans who had migrated to North America from settlements in Russia.

on the ground that many Russian Germans could speak English and that the Scandinavian-Americans and Italian-Americans in the state had not asked for religious services in their native languages. Vigilantes tarred and feathered several Mennonites in Kansas and shaved off the beards of others. In Oklahoma, mobs attacked one Mennonite twice, the second time after he had given seventy-five dollars to war relief as a substitute for buying Liberty Bonds. Terrorists ransacked his home and smeared it inside and out with yellow paint, removed his clothing and beat him twelve times with a leather strap, then tarred him with roofing compound.

Near the end of October 1917, a group of masked men abducted Herbert S. Bigelow, pacifist minister of the People's Church in Cincinnati. Bigelow had been an ally of Secretary of War Newton D. Baker in the fight against Ohio machine politicians and later became a socialist. He had delivered a sermon in which he asked God to cause all the warring armies to cast down their arms. In early October, agents of the Department of Justice went through his church looking for subversive material. According to his account, as he was about to address a socialist antiwar meeting a few weeks later, he was seized outside the hall, handcuffed, and driven a considerable distance in an automobile with a bag over his head. When the car stopped, the bag was removed and he was tied to a tree. In the moonlight, he could see about twenty-five to forty men wearing white masks and white aprons or skirts surrounding him. His overcoat, jacket, vest, and suspenders were removed. Someone said, "In the Name of the women and children of Belgium and France lay on," and a man who had stepped forward with a blacksnake whip lashed him repeatedly. Another man began cutting locks from the top and front of his hair, and crude oil was poured on his head. He was then permitted to dress, but the leader of his tormentors told him to wait ten minutes after the mob had left and warned him to be out of Cincinnati in thirty-six hours and to stay out for the rest of the war. Attacks like these went on for months.

Like pacifists and German-Americans, striking workers and their leaders were attacked by vigilante mobs. In the summer of 1917 during a labor dispute in the Bisbee, Arizona, copper field, mining-company officials and other citizens arranged to expel strikers from the area. On July 12, the sheriff and about 2,000 armed men rounded up 1,186 mine workers, several of them members of the IWW, and deported them by train to Columbus, New Mexico. When local officials refused to accept the deportees, the train carried them to the desert town of Hermanas, where their guards abandoned them with little food and water. Two days later, army personnel rescued them and brought them back to Columbus, where the federal government maintained them until mid-September.

Less than three weeks after the Bisbee deportation, masked men dragged a one-eyed, part Native-American IWW agitator, Frank Little, from his room in Butte, Montana. Miners were striking the Butte copper fields, and Little had not only encouraged them to stay off the job but had made antiwar remarks at a rally, allegedly calling American soldiers "uniformed scabs." The intruders tied Little, who was crippled by rheumatism and a broken leg, to a car and dragged him through the streets, scraping the skin off his kneecaps. Then they hanged him from a trestle with a sign pinned to his underwear: "Others take notice. First and last warning. 3-7-77" (an old vigilante symbol representing the dimensions of a grave). Several Montana newspapers approved the lynching as a patriotic act.

The wartime hysteria, of which these are only some of the reported incidents, arose from many causes: hatred of foreigners, whom older-stock Americans believed were taking over jobs and neighborhoods and changing American institutions; disturbing wartime economic and social dislocations; anxiety over the fate of loved ones in the armed forces; spy scares; and rumors that German-Americans were planning insurrection, mixing ground glass into flour, dynamiting factories, and spreading the influenza germs that swept Americans away by the thousands during the epidemic of 1918. All these contributed to a fanatic nationalism; so did the reality of prewar German espionage and sabotage, the fact that Germans were killing and wounding Americans on battlefields and in submarine attacks, and the refusal of some German- and Austrian-Americans, even after war was declared, to hide their sympathy toward the Central Powers.

Portrayals of German-Americans in the media contributed to the hysteria by making them seem responsible for actions of the German state. Thus, a two-panel cartoon by A. B. Walker appeared in *Life* during the April 1918 wave of violence. It showed a munitions plant exploding, a German-American running away from it, and then an army firing squad shooting down German-Americans. The caption read: "We Would Have Less of This. . . . If We Had More of This."

The hysteria of 1918 exemplifies the way many groups of Americans turned the war into an opportunity for gain. War provided people with seemingly legitimate reasons for attacking personal or social enemies. Members of ethnic groups who had long resented German-Americans were able to assail them, to harass and silence them in the name of patriotism. This ethnic hostility helps explain why one state council of defense continued its ban on the German language and German textbooks even after the war was over.

Members of the business community used the emergency to regain some of the power and influence they had lost before the war to farmers, workers, and reformers. Their instruments were often voluntary war agencies, particularly the councils of defense, on which they usually outnumbered all other groups. In the fall of 1917, a Wisconsin farmer complained to the assistant secretary of agriculture that in selecting bodies that were supposed to represent the nation, such as defense councils and public safety commissions, "we have surrounded the farmers with the same class of people who have incurred their ill will in former times, the bankers, lawyers, professional politicians, representatives of a business system . . . who now under the guise of Patriotism are trying to ram down the farmers' throats things they hardly dared before."

The farmer's letter described a pattern that appeared in many communities throughout the country. George Creel had it in mind when he urged President Wilson, a few days before the Armistice, to begin demobilizing the Council of National Defense "so that the chauvinistic, reactionary state organizations may be put out of business."

Participating in attacks on internal enemies provided other, less-tangible advantages to certain people. It gave them a chance to experience feelings of fraternity and power or to act out fantasies that could not be performed in ordinary circumstances—the fantasies of those who undressed, lashed, tarred, feathered, and murdered people, humiliated them and painted them yellow, clipped locks from their hair, poured oil on their heads, or made them kiss flags.

War hysteria afforded some the chance to avert suspicion of their own loyalty. A few German-Americans tried to demonstrate fidelity to the United States by joining or assisting mobs that harassed people of their own national origin. In American courtrooms, respectable citizens expressed patriotism by convicting those accused of violating the Espionage and Sedition acts. Judge Charles F. Amidon declared,

> For the first six months after June 15, 1917, I tried war cases before jurymen who were candid, sober, intelligent businessmen, whom I had known for thirty years, and who under ordinary circumstances would have had the highest respect for my declarations of law, but during that period they looked back into my eyes with the savagery of wild animals, saying by their manner, "Away with this twiddling, let us get at him." Men believed during that period that the only verdict in a war case, which could show loyalty, was a verdict of guilty.

A juror who helped acquit leaders of the mob that lynched Robert Prager shouted, "Well, I guess nobody can say we aren't loyal now."

Overshadowing all these reasons for war hysteria were the actions of the federal government, particularly of the executive branch. The Wilson administration encouraged fanaticism by its support of superpatriotic extremists, by its own repressive actions, and by its failure to speak out firmly, consistently, and in a timely manner against attacks on individual freedom. It promoted divisiveness and suspicion with propaganda in CPI posters: "Have You Met This Kaiserite?" which urged traveling salesmen to send the Department of Justice names of any disloyal people they encountered; and "Spies and Lies," which warned that German agents were everywhere and asked its viewers to report anyone spreading pessimistic stories. As Assistant Attorney General O'Brian observed, the constant patriotic agitation of Liberty Loan speakers, Four-Minute Men, and others "worked the whole country up to a pitch of intensive patriotism, resulting in instinctive aversion toward anyone even under suspicion for disloyalty."

Yet it should not be imagined that the president and his assistants had some detailed master scheme for turning people into bigots. The situation was more complex. The Wilson administration was deeply ambivalent toward repression. It contained hard-liners like Burleson along with advocates of free expression like O'Brian. It was constantly pressured by outside fanatic elements whose fanaticism the war and administration policies had helped create. A repeated theme of persons who wanted to crack down on dissenters or German-Americans was that if the government did not do it lawfully, mobs would do it outside the law. Administration leaders began to feel that they must take command of the campaign against dissent or there would be even more restrictive legislation in Congress, more agitation in the states, more mobbings, and more lynchings by local groups. They realized that Republicans, who never allowed the war to interfere with politics any more than Democrats did, were using war hysteria as a means for achieving power—for instance, by taking control of such volunteer organizations as the councils of defense that could be used as a power base. And they knew that if Democrats appeared too soft on dissenters, men with tougher-sounding policies would likely replace them at the next election.

The Wilson administration's two-sided attitude toward repression expressed itself in a tendency to take differing and often contradictory stands. Thus, we find Attorney General Thomas W. Gregory—whose department deputized the American Protective League, who encouraged Americans to act as "voluntary detectives" and informers, and who wanted a federal sedition act because the Espionage Act did not cover casual, impulsive, or truthful remarks—publicly stating in 1918 that the rights of free speech, the press, assembly, and petition continued to exist in wartime and telling federal attorneys that the Sedition Act should not be used for personal vendettas or to suppress honest, legitimate discussion of government policies. Though Gregory's assistants O'Brian and Alfred Bettman believed in freedom of expression, they sanctioned the semiofficial law-lessness of the APL, and neither they nor anyone else in the Justice Department did much to curb the excesses of federal district attorneys until just before the Armistice.

Wilson's own words and actions typified the ambivalence of his admin-istration. As early as 1914 he had expressed the fear that America might have to take defense measures fatal to her ideals and form of government. He had anticipated the wartime conflict between ethnic groups, telling the German ambassador that the United States would have to remain neutral lest "our mixed populations would wage war on each other." Now he warned Americans not to copy Germany by ignoring the "sacred obliga-tions of the law" or by attacking organizations that would produce anarchy by anarchic methods. It was "childish," he thought, to prevent school children from learning German. Privately, he expressed deep concern about "the treatment of people whose offense is merely one of opinion."

The president spoke out against mobs and tried to moderate particular efforts at censorship. He urged Postmaster General Burleson to act with "utmost caution and liberality in all our censorship." When the Post Office sought to exclude the *Milwaukee Leader* from the mails, Wilson remarked that there was "a wide margin of judgment here and I think that doubt ought always to be resolved in favor of the utmost freedom of speech." He intervened to release the *Nation* and the *World Tomorrow* when Bur-leson's department excluded them from the mails, and he disagreed with Burleson's view that the government should take drastic action against the People's Council of America for Democracy and Peace, an antiwar organi-zation—so long as it stayed within the law. When legislation was proposed that would have subjected civilians to court-martial and possibly a death penalty if they published anything that endangered the success of the military forces, Wilson successfully opposed it.

Yet it should also be noted that Wilson's criticisms of lynching mobs were infrequent and belated, that he encouraged the attorney general to initiate treason proceedings against publications, and that, in the Rose Pastor Stokes case, he urged the Justice Department to see if there were a way of punishing the editor who allowed Mrs. Stoke's letter to be pub-lished. When the Espionage Act was being framed, he tried unsuccessfully to have a provision added giving broad censorship powers directly to the president; he favored a peacetime sedition act (possibly to head off more stringent legislation); and in 1920, long after the fighting was over, he pocket vetoed a bill that would have abolished the Espionage and Sedition acts.

Wilson assured a journalist that he would not allow the Espionage Act to be used to shield himself and his actions against criticism and that he would regret losing the benefit of "patriotic and intelligent" criticism. Nevertheless, he lashed out at progressives who spoke against the war, accusing them of being tools of the German Empire, and expressed deep personal hostility toward individuals who attacked his war policies. After reading a newspaper article about a Mr. Hannis Taylor, who argued that it was unconstitutional to send draftees or militia overseas and who warned that the president's proclamation calling the militia into service was leading the United States toward military dictatorship, Wilson wrote the attorney general, "Do you think there is anything we could do to this wretched creature . . . or is he too small game to waste powder on?" The president described Postmaster General Burleson as "inclined to be most conservative" in censoring the mails. But if Burleson was conservative, what did Wilson suppose an aggressive censor would have done?

Wilson told the editor of the *Masses*, Max Eastman, that "a line must be drawn" and that his administration was trying to draw it "without fear or favor or prejudice." Yet the actions of the Wilson administration in specific censorship cases were sometimes capricious or affected by political considerations. The left-wing *Milwaukee Leader* did not get its mailing privileges back until Warren Harding took office, but Theodore Roosevelt and conservative dissenters were permitted to deliver public attacks against the government throughout the war.

What saved the *World Tomorrow* and the *Nation*? In the case of the former, it was intercession with the president by John Nevin Sayre, an Episcopal minister and brother of Wilson's son-in-law. And while the *Nation* did not have that kind of immediate access to the chief executive, the impeccably upper-class background of its editor, Oswald Garrison Villard, placed it in a different category from ordinary left-wing magazines.

In 1926 George Creel remarked that at the height of the war, the president had been against free speech. The former CPI chairman oversimplified the situation, ignoring the special status of powerful and influential people who remained free to say what they wished. On the whole, though, Creel was essentially right, because to manage a divided nation in a total war, Wilson felt compelled to follow Lincoln and John Adams and limit the freedom of ordinary Americans to dissent. A time of war, the president said, "must be regarded as wholly exceptional," and it was legitimate to regard things "which would in ordinary circumstances be innocent as very dangerous to the public welfare."

What of Creel's own agency, the organization charged with most responsibility for managing public views? There one finds the same inconsistent attitude toward repression. The CPI warned its Four-Minute Men against preaching hatred; yet it encouraged them to breed fear among civilians and warned them that they could not accomplish their objectives if they always talked on a highly ethical level. Creel himself urged Wilson, just before the Armistice, to put the state councils of defense out of business. Yet in August 1917, Creel wrote Lionel Moses of Minneapolis and urged him to incite citizens against the antiwar People's Council, which was planning a convention in that city. The People's Council was made up of "traitors and fools," Creel declared, "and we are fighting it to the death. . . . Have patriotic societies and civic organizations pass resolutions condemning the People's Council as pro-German and disloyal. . . . Get a good committee together to go around and see all the newspapers and see to it that they get the point of view and action that I am giving you now." Creel added, "Tear this letter up."

Observing the actions and the words of the Wilson administration, as it sought to control the way Americans thought, felt, and acted, we see that it employed and encouraged repression with some trepidation and in response to pressures partly generated by itself. From the viewpoint of jailed Wobblies, fired professors, terrorized pacifists, harassed German-Americans, silenced dissenters, and those who valued the right to protest, even in extraordinary times, it made little difference whether the administration felt concern about what it was doing to personal liberties; to such persons, wartime repression was just a disaster.

Still, one has to ask how the Wilson administration could have followed a different course. If the president had been more deeply committed to preserving free speech, his government might have treated dissenters less harshly. But that would have made him and the Democratic party more vulnerable to attacks from those infected with the "reactionary chauvinism" that war helped elicit. It might also have allowed opposition to the war to spread, undermining military production and troop morale. The administration faced a common problem of democracies: how much liberty is it possible to tolerate while successfully waging a large-scale war?

The United States government was never able to stamp out all disaffection, but what it did proved good enough. Enthusiasm for the crusade spread and intensified among the populace; most dissidents learned to keep silent; and the American people became mentally prepared, not only to serve on the battlefronts but also to join in the massive economic mobilization that the Great War demanded of belligerent nations.

3 ‖ The Managed Economy: Creating the Regulatory System

Of all the belligerents in the First World War, the United States was potentially the best suited for waging the kind of struggle the Great War turned out to be—an extended conflict in which economic might counted at least as much as military skill. America was then the foremost industrial and financial power, with resources that seemed endless. It had a highly developed transportation network, a newly reformed national banking system, productive farms, armies of workers, and a large pool of managers experienced in mass production. Yet having devoted most of its energies since the Civil War to internal development, it was unprepared for rapid mobilization and as it tried to convert its economy for war suffered terrible dislocations. Out of this turmoil, a centralized war capitalism emerged, run chiefly by the representatives of the country's great corporations under the aegis of the United States government.

The Economic Crisis

From the experience of the European belligerents, the United States government had ample warning of how difficult its own mobilization might be, and it began some sketchy preparations for a war economy. Nevertheless, before April 6, 1917, the Wilson administration lacked the resources, the knowledge, and the legal authority to manage conversion from peace to war. There was no machinery for determining accurately what the Allies, the armed services, and Allied and American civilians would need; no satisfactory method of estimating what American industry could produce; no adequate procedure for apportioning food, fuel, raw materials and finished products; no effective mechanism for coordinating the country's transportation systems; and no agency to regulate prices. With the nation's economy operating near its capacity and in the absence of effective controls, an economic crisis developed that was as damaging, in its way, as the First Battle of Bull Run or the burning of Washington in the War of 1812.

Trouble took innumerable forms. Military procurement, especially in the War Department, was chaotic. While the navy had been growing since the 1880s and had developed fairly efficient methods for buying what it needed, the army, which expanded in a few months from less than two hundred thousand men to over two million, was ill-prepared for the change. The War Department had several virtually independent systems for buying, financing, storing, and transporting military goods, each serving a particular bureau, such as the Medical Department, the Ordnance Department, and the Quartermaster Corps. These bureaus, old and entrenched, with powerful friends in Congress, had fought off attempts at reform for years and were in no hurry to adapt to the buildup of 1917. The organization of the bureaus made it difficult to match their needs to the factories producing what they wanted to buy. The army's supply systems were arranged by military functions—for instance, medical supplies, ordnance supplies, engineering supplies—while American industries were organized by commodities—such as pencils, steel, and shoes. The bureaus, therefore, had to reorganize their supply networks to fit the economy, which they refused to do; or the economy had to reorganize to fit the War Department, which was impossible; or a way had to be found to harmonize the two systems.

The effects of this antiquated supply system appeared almost at once. The army continued to deal with companies that had supplied its bureaus in peacetime, placing most of its orders in the Northeast, which could not meet the demand. Since the bureaus had only fragmentary ideas of what they needed, vague information about what they already owned, and no arrangement to determine which agency should buy which materials and in what order, they competed for goods with one another and with the navy, the Allies, and civilians. Buying everything they supposed they might require, commandeering what they could not purchase, the army bureaus created economic bedlam. A biographer of Secretary of War Baker describes how the secretary went down to the cellar of the War Department one day and found the corridors stacked with typewriters from floor to ceiling. The adjutant general, Baker learned, had tried to purchase every available typewriter in the United States, fearing that, otherwise, the surgeon general, the Treasury Department or the navy would have bought them all.

The War Department's frantic orders and other military purchases poured billions of dollars into the economy. Production could not begin to satisfy demand. Prices soared and fluctuated. Believing that whatever they bought they could sell later at higher prices, speculators started to hoard. Producers and the government could only guess what future costs might be. As prices rose, workers clamored for higher wages and often received them, because in many trades the need for labor far exceeded the number of workmen available. Besides, the government was reimbursing war contractors for labor costs, so raises could be charged to Uncle Sam. Businessmen who needed highly skilled workers to finish contracts sometimes bid their wages to levels unimagined in ordinary times. Yet many other workers found it hard to keep up with rising living costs and continued to endure harsh working conditions. Despite pay increases, the number of stoppages over wages or hours rose from 2,036 in 1916 to 2,268 the following year. Some of these strikes were in industries like copper and lumber, which produced vital war materials.

As strikes, inflation, and the army's procurement system snarled mobilization, the national transportation network began to collapse. The railroad network had been in poor condition before 1917, its equipment overworked and undermaintained, and its finances drained for decades by speculators. When railroad companies argued that they needed to increase freight rates to rebuild and repair and to pay increasing labor costs, shippers complained that the roads were making plenty of money on war traffic. State and federal regulatory commissions denied most of the carriers' requests. The carriers attempted to coordinate shipments, passing freight from one railway line to another, but some of their local managers, imbued with competitive spirit, refused to send traffic to other companies. When the railroads tried to unify and rationalize their industry, the United States attorney general threatened to prosecute them for violating antitrust laws. They sought to borrow half a billion dollars, but the Treasury stopped them because it wanted investors to spend their money on government war loans, not railroad bonds.

War traffic began to overwhelm the railroad system. Because the American Expeditionary Force was embarking from ports in the Northeast, where the services sent most of their orders, decaying lines of a single region had to carry millions of troops, vast quantities of military supplies, the materials needed for war production, and regular civilian passengers and freight. A shortage of ships added more burdens, for with U-boats threatening to sink vessels faster than they could be built, coastal ships were sent into transatlantic service, leaving their cargoes to be carried in railroad cars.

After a while, the shortage of vessels began to hurt the roads in a second and even more crippling way. Since there were not enough ships to carry goods from the docks as fast as trains delivered them, terminals at East Coast ports began to fill up, warehouses first, then sidings, until it became necessary to empty freight cars in fields ten miles inland. Soon it was impossible to unload anywhere near the coast.

Meanwhile, trains continued rolling in from the west. Some Allied purchasers paid for part of their orders as soon as the orders were loaded. This led companies in the Midwest to place shipments in freight cars and, regardless of whether the cars were filled or not, to dispatch them toward the great pile-up. The armed forces and the federal shipbuilding agencies, which enjoyed special transportation priorities, continued to send cargoes into congested areas, such as the Hog Island shipyard under construction on the Delaware River. At one time, eleven miles of freight cars were lined up at the entrance to the yard, all containing materials from which the yard itself was supposed to be built. The cars could not go inside because there were no tracks on the construction site and no facilities for unloading.

Eventually congestion became so bad that traffic intended for eastern ports was blocked all the way to the west of Chicago. It was hopeless to attempt to unload freight cars, because to separate out any particular car required moving the ones in front or in back of it on completely jammed tracks. Soon no empty cars could be found to carry new cargoes. Then, an extraordinary winter descended, numbing the hands of railroad workers, freezing switches, and blocking tracks with snowdrifts.

With the nation's railroad system slowing to a halt, munitions could not be moved. A coal shortage, resulting chiefly from the absence of available fuel cars, threatened war production east of the Mississippi. Local authorities began to confiscate coal moving through their towns. With transportation blocked and the nation compelled to feed its new expeditionary force as well as the Allies, food supplies in some places began to run out, and American civilians faced the prospect of going hungry in the first winter of war.

Congress investigated. Senator George E. Chamberlain of Oregon exclaimed that the American military establishment had almost stopped functioning. Republicans and even a few Democrats clamored for a coalition war cabinet. Not even Secretary Baker's statement that "this War is being fought in every factory, every workshop, every home in the country by those marvelously subtle processes of modern scientific achievement by which we are all coordinated" prevented people from thinking that mobilization was a shambles.

Resolving the Problems

Yet while critics complained and patriots lamented, the federal government was taking steps that would eventually resolve the crisis. Building on a foundation laid down before the war, people in Washington were putting together a structure of mobilization agencies and developing ways to increase the output of essential goods and services, to improve distribution, and to channel resources from civilian to war-related enterprises.

By the summer of 1917, government officials realized problems would soon develop with the nation's food and fuel supplies. In August, Congress provided an instrument for solving those problems when it passed the Lever Food and Fuel Control Act, one of the most sweeping grants of power in American history. The Lever Act authorized the president to regulate the output, distribution, and price of food and to control every product, including fuel, that was used in food production. Wilson then established two regulatory agencies: a Fuel Administration, to which he named as director Harry Garfield, the president of Williams College; and a Food Administration, headed by Herbert Hoover, the well known engineer who had directed a relief program for the Belgians and helped Americans get back from Europe at the beginning of the war.

The two agencies took steps to expand output while restricting civilian consumption. To stimulate production of pork and wheat, Hoover's agency paid premium prices. It bought up the entire 1918 sugar crops of the United States and Cuba and limited the amount of sugar civilians could purchase. To stimulate voluntary cooperation, which Hoover, for a variety of reasons, considered far more desirable than coercion, the Food Administration instituted programs of voluntary meatless and wheatless days, sending half a million people out to secure pledges from American housewives that they would conserve food. The twenty million people who signed these pledges received buttons and window stickers to show that they were cooperating. The Fuel Administration declared fuel "holidays," shutting down manufacturing plants for five days starting January 18 and ordering all but a few essential businesses to burn no fuel on nine consecutive Mondays. It introduced daylight saving time to save energy and set coal prices so high that even inefficient mining operations became profitable. As a result of these measures and of an ample harvest in 1918, the output of coal and agricultural products gradually caught up with demand.

To mobilize workers effectively, the administration had to solve some extremely complex problems. Since the American war machine depended on a balance of military force and economic power, the government had to find a way to hold thousands of men in essential civilian war jobs while sending others overseas to fight. It had to keep strikes from disrupting production without encouraging inflationary wage increases. In certain war industries, it had to reconcile the demands of workers—for shorter hours, improved working conditions, and the right to form unions—with the refusal of managers to loosen their control of the work force. Workers, management, and sometimes both ignored appeals to patriotism. Yet patriotism sometimes proved too powerful—when employees and managers abandoned vital civilian jobs to put on the uniform.

Gradually, the federal government developed a series of organizations to work out these problems. The armed services and civilian agencies, such as the Fuel Administration, created special bureaus to handle labor issues. A United States Employment Service, part of the Department of Labor, helped unskilled workers find jobs in war industries, while the Selective Service System channeled men into essential occupations by offering them draft exemptions. A presidential mediation commission traveled around the country seeking to forestall or resolve labor disputes in vital industries. A National War Labor Board, composed of five labor members and five employer representatives, with President William Howard Taft and attorney Frank P. Walsh as co-chairmen, coordinated the work of federal mediators, set minimum wages based on living costs and gender, and served as a court of last resort in labor disputes.

The new labor boards, created independently of one another, at first proceeded in uncoordinated ways, sometimes aggravating the labor problems they were trying to solve. During a munitions strike in the fall of 1917, mediators appeared from four agencies, each with different orders and a different policy for settling the dispute. By making uneven awards and enforcing varying labor standards, the federal labor agencies induced workers to move from job to job for higher pay, and they encouraged unions to raise their wage demands, since a union in one plant could insist on the higher wage scale that a board had approved for a similar plant somewhere else. In an effort to cope with these difficulties, the administration brought officials from the major war agencies into a War Labor Policies Board, which devised standard labor policies for every region of the country.

Ordinarily, the federal government tried to resolve labor disputes with an indirect approach, relying on mediation and gentle persuasion. It cooperated with conservative, pro-war labor leaders, offering, in return for their support, improved pay and working conditions and the right to recruit war workers into those unions. By basing profits on costs, including the cost of labor, government contracts encouraged employers to raise workers' pay. But where unions and workers were uncooperative or militant and threatened the supply of war materials, the administration used tougher methods. During the copper strike in the summer of 1917, federal agents raided the halls and headquarters of the Industrial Workers of the World; seized files, office equipment, and literature; and arrested hundreds of leaders. Federal troops entered mining fields in Montana and

Arizona to halt labor unrest. When a labor dispute in the Pacific Northwest threatened the supply of lumber for warplanes and ships, the army enlisted forest workers in a federal "company union"—the Loyal Legion of Loggers and Lumbermen—and sent out patrols to keep American Federation of Labor and IWW organizers away from the lumber camps. President Wilson broke a strike at an arms factory in Bridgeport, Connecticut, by telling the strikers that if they did not return to work, he would see to it that they lost their occupational exemptions.

While it attempted to mobilize the country's workers, the federal government struggled to keep the country's transportation functioning, particularly the ocean-shipping and railroad systems. In 1916 Congress had authorized the executive branch to create a United States Shipping Board, which the Wilson administration used to commandeer German, Dutch, and partly finished American ships and to take control of certain privately owned American oceangoing vessels. The government established a War Risk Insurance Bureau in the Treasury Department to supply shippers with public insurance and set up an Emergency Fleet Corporation to build shipyards and merchantmen. It employed hundreds of experts, under the direction of the dean of the Harvard Graduate School of Business Administration, to analyze the operations of the nation's ocean shipping system and to recommend improved methods for loading, unloading, and transporting cargoes.

At the height of the transportation crisis, in December 1917, the federal government took control of most of the country's railroads, together with railroad terminals; warehouses; and express, sleeping-car, and telegraph companies. The president created a Railroad Administration, named his son-in-law William G. McAdoo, the secretary of the Treasury, as its director-general, and placed it in charge of nearly four hundred thousand miles of track, the property of nearly three thousand companies.

McAdoo's agency enjoyed several advantages over the private railway companies. While they were chronically short of funds to replace aging equipment, rebuild their lines, and pay wages to match increases in the cost of living, the Railroad Administration could draw on the credit of the United States. Antitrust laws, traditions of competition, and obligations to banks or to other creditors did not hamper it. It could countermand shipping orders of government agencies. Most important, it could unify the roads into a coordinated network.

McAdoo was thorough and ruthless. He removed hundreds of railroad executives and placed the rest in two chains of command, one corporate and one federal, both under his authority. Stationing Railroad Administration representatives in the War Department and other government pur-

chasing agencies, he overrode their priorities and saw to it that no one
shipped into congested areas without assurance that freight cars could be
unloaded at their destination.

Under the Railroad Administration's guidance, the American railroad
system became far more efficient than it had been before the war. Government managers increased the number of cars each locomotive had to haul,
enlarged the loads of the cars, and held up small shipments until they
could be combined into carload-size lots. They suspended unessential
traffic, discontinuing four hundred passenger trains during the agency's
first two weeks. Since the Railroad Administration did not have to confine
its traffic to any particular company's lines, it shipped by the shortest
routes available, and it sent new locomotives where it judged they were
needed, regardless of which company had ordered them. It standardized
freight cars and established coal shipment zones to prevent people from
hauling coal from distant mine fields when there were mines nearby.

McAdoo's organization stopped railroads from competing for freight
business and removed agents who had once been paid to solicit freight
orders. It halted railroad advertising, made terminals serve more than one
line, and consolidated ticket offices. To prevent labor problems from
erupting, the government gave railroad employees a $300,000,000 pay
raise. Meanwhile, the government provided railroads with capital and
guaranteed them a return based on their prewar profits.

The people who owned and managed the railroads had mixed feelings
about all these changes. Some owners complained that the government
overworked their equipment and undermaintained it and paid them less to
use their property than it should have. Managers who stayed on to work
for the Railroad Administration took severe cuts in pay, while redundant
executives were naturally displeased at having to look for other jobs. Yet
the significant facts are that many private managers did stay on, that they
dominated the government railroad agency, and that as executives of the
Railroad Administration they did not have to worry about being prosecuted for anticompetitive practices and were no longer restricted by state
regulatory commissions. The reforms instituted during the war were
largely those that owners and high-level executives of railroads had wanted
to effect before federal control.

The Railroad Administration encouraged business self-regulation, a
common wartime phenomenon in which corporations, with government
sanction, used the war to reorganize their industries. This phenomenon
generally occurred in industries dominated by a few powerful companies.
Industries that were fragmented and weak had much less leverage with the
administration, which tended to run them with tighter controls.

The actions of the Railroad Administration produced winners and losers. The winners, at least while the war lasted, were the government, which was able to move its goods more efficiently; reformers, including McAdoo, who wanted to demonstrate the efficiency of a federally controlled railroad system; railroad workers who received higher wages; and those railroad managers and owners who abhorred needless competition, hated regulation by unsympathetic government agencies, and were able to emerge from the war with new images—as public servants. The chief losers were the state railroad commissions, now all but removed from the system; people who still clung to the competitive ethic; travelers inconvenienced by the loss of some routes and ticket offices; and shippers who had to pay higher freight costs and lost their ability to influence freight rates through state regulatory agencies. Yet the shippers also benefited from having a railroad system that worked.

Although several of the complaints of the Railroad Administration's critics were valid, the agency performed its task effectively. Key indicators—net ton miles, average carload size, and the net tonnage of an average train load—all showed that the American railroad system was working more efficiently in 1918 than in the days before government control. By the end of the war, there was actually a surplus of freight cars. McAdoo's organization turned out to be one of the most successful accomplishments of the American war government.

While he struggled to restore the country's railway system, McAdoo, in his role as secretary of the Treasury, faced another daunting task—developing ways to finance the war. What made this second task so challenging was the inability of the American economy, already working almost at capacity in April 1917, to supply everyone who made demands on it—including the American armed forces, the Allies, and American civilian consumers. One result, as we have seen, was higher prices.

The Treasury Department used two methods to limit demand from civilians: taxing them and borrowing their surplus funds. For political reasons, the government ended up paying for the war mostly by borrowing. Originally, McAdoo hoped taxes would cover about half the war costs, but when pressure groups demanded that Congress keep taxes down or shift them to other segments of the population, the legislators responded by raising only enough taxes to finance less than a third of the expense of the war.

Even so, taxes rose significantly and were generally progressive; that is, they took higher percentages of income from those in higher income brackets. The 77.7 percent of American taxpayers whose income was less than $3,000 in fiscal year 1918 contributed less than 3.6 percent of all individual tax receipts and paid the federal government no more than 4 percent of their income.

Table 1   Tax Rate by Income Group

Year	$2,000	$5,000	$10,000	$50,000	$100,000	$1,000,000
1915	—	0.4	0.7	1.5	2.5	6.0
1917	1.0	2.4	4.0	10.4	16.2	47.5
1918	3.0	4.8	9.5	22.3	35.2	70.3

Source: *Historical Statistics of the United States: Colonial Times to 1957* (Washington, D.C.: U.S. Bureau of the Census, 1960), 716.

Efforts to absorb buying power by borrowing from the public were only partly successful. The Treasury Department sold lower-income citizens several hundred million dollars' worth of inexpensive war stamps and small-denomination certificates, but its Liberty Bond drives, for all their spectacular publicity (and all the grief they caused to those who were pressured to contribute), actually increased the amount of money available for civilian spending.

To keep financing costs down and to protect savings banks from competition by the Treasury, the government paid low interest rates on Liberty Bonds. Yet if interest on the bonds was very low, people might not buy enough of them to fill national quotas. To forestall this embarrassing result, the Treasury Department and the Federal Reserve Board made it easy for banks to lend money to bond purchasers. These loans built up inflationary pressures by increasing the money supply. The Treasury sold billions of dollars' worth of bonds and certificates to the banks, which used them as a basis for lending more money, and allowed the borrowers to use Liberty Bonds as collateral for loans. By making certain bonds tax exempt, it enabled wealthy persons to reduce their taxes, liberating even more money that could be used to bid for goods and services against the armed forces, the Allies, and other organizations directly involved in the war effort.

With all their inconsistencies and shortcomings, these financing programs had considerable long-term importance. Not only did they help the country pay for its part in World War I, but they also demonstrated how the United States government could draw on new powers that the income-tax amendment and Federal Reserve Act of 1913 had given it to shape the structure of a war economy and finance a multibillion dollar military buildup.

The keystone of that structure, the principal coordinating body for the American economy, was the War Industries Board (WIB). Composed of representatives of most of the nation's major economic interest groups and dominated by industrialists, this agency matched the needs of the armed forces and other war organizations, the Allies, and American consumers with the output of the nation's producers. It culminated a series of efforts begun years before the war to link the armed forces with business and the professions and to construct a "scientifically" managed mobilization system.*

The Wilson administration had fostered those early efforts as part of a prewar preparedness campaign. In 1915 Secretary of the Navy Josephus Daniels brought several members of industrial and engineering societies together in a Naval Consulting Board, which evaluated technology with military applications. Its Industrial Preparedness Committee, directed by Howard E. Coffin, an auto company executive and an energetic advocate of scientifically managed war mobilization, drew up lists of factories that could turn out military products. With the assistance of the army chief of staff, another team headed by Hollis Godfrey (the president of Drexel University and a promoter of scientific management in industry and defense) conducted an analysis of national resources with potential military applications. Godfrey's group sought to determine the ingredients of every product needed for war, where these ingredients could be found, and how they could be assembled. In addition, at the request of President Wilson, the National Academy of Sciences created a National Research Council to coordinate national-security research by scientists and engineers. In 1916 the administration submitted and Congress approved a section of an army appropriation bill that incorporated ideas of Wood, Godfrey, Coffin, and others, providing for the Council of National Defense (CND), the industrial mobilization agency with which the country began the war.

This council consisted of six cabinet members under the leadership of the secretary of war. Attached to it was a National Defense Advisory Commission (NDAC) made up of representatives of business, the professions, and organized labor, who, together with the CND, were expected to advise the president on ways to mobilize the economy. Each branch of the NDAC developed staffs of experts, and when the United States entered the war, the CND also developed "cooperative committees," members of which were recruited from the leading producers. Participants on these committees were called dollar-a-year men because the government paid them a nominal sum while their companies paid their regular salaries. They determined for their industries what government purchasers needed to buy, recommended suppliers, and suggested what the government should pay.

When the mobilization crisis began, the Council of National Defense,

*Scientific management referred to an effort to make business more efficient through systematic analysis and reorganization, including the reorganization of jobs.

its advisory commission, and the cooperative committees struggled to put military purchasing in order. The CND tried to find out what the army and other government agencies really needed to buy and attempted to stop them from competing with one another for supplies and from overloading some companies with orders while ignoring others. It sought to end mindless commandeering. The results were uninspiring. Army bureaus looked on the CND and its offshoots as an enemy force. Since the Council of National Defense and its subsidiaries did not have legal authority to make purchases and the armed forces did, the civilians were forced to rely on appeals to common sense and other mild forms of persuasion.

As they battled with the military services, the Council of National Defense and its branches struggled to bring producers under control, to replace the market system with a system of allocation. The CND tried to get companies to sell standard products at standard prices to the people who needed them most to fight the war. This was extremely difficult, particularly in decentralized industries, like lumber, where every mill sold to whoever paid it the highest price. Only a minority of zinc producers even cared to bid on war contracts, and it took over six months to bring most cotton mills into line. Some mill owners did not like the idea of working with competitors, not just because they were attached to the market system, but also because they were afraid the government might prosecute them under the antitrust laws. In addition, some companies had bad memories of paying graft for military contracts before the war or did not want government inspectors roaming through their plants; some disliked the government's regulations, feared they would not be paid on time, or worried that the government would cancel their contracts as soon as the fighting stopped.

Then there were problems with the Council of National Defense itself, particularly with the cooperative committees. The dollar-a-year men who worked for those committees were in a position to award highly profitable contracts to their own companies, a point not lost on competitors. After businesses that failed to receive war contracts complained about the possibility of conflict of interest, Congress wrote criminal penalties into the Lever Act, making it unlawful for persons working for the government to solicit contracts in which they had an interest. Cooperative committee members were already afraid of prosecution for allocating contracts without legal authority, and after the Lever Act passed, some of them quit. The chief of the committee on army and navy artillery, Samuel Vauclain, vice president of the Baldwin Locomotive company, wrote, "I have been up against several propositions in my time but I do not want to get up against the penitentiary."

These allegations and fears, and particularly the CND's lack of power, led members of its advisory committee to call for a stronger mobilization system. The result was the War Industries Board, which the Council of National Defense, with President Wilson's approval, created in July 1917. The new agency began to supervise all of the nation's industrial production for war. It did away with the cooperative committees and included on its governing board representatives of the army and navy who could tell their services what the WIB was doing and give them a voice in its activities.

It took months before the WIB began to function effectively. At times, its officials were contemptuous of the military, while army officers showed no more love for War Industries Board executives than they had for the civilians who worked for the Council of National Defense. Sometimes the War Department acted as if the WIB did not exist. It proceeded independently on a large munitions contract, a huge storage and embarkation base, and dozens of other projects that affected the economy in significant ways and kept seizing supplies and factories after the WIB asked it to stop. Despite persistent requests from board members and urgent pleas from the American Expeditionary Force, it refused to abandon its tedious, and in the board's opinion, unnecessary procurement procedures. Sometimes the War Department refused even to tell the War Industries Board what the army needed.

Under these circumstances, it is not surprising that WIB chairmen found their job unbearable or that one of them, industrialist Frank A. Scott, became physically and spiritually too ill to go on. For a while, Daniel Willard, president of the Baltimore and Ohio Railroad, tried to handle the job, but in the black days of mid-December 1917, he quit after blasting the War Department before a Senate committee and insisting that one man should have authority to control all war purchasing. For the next three weeks, the War Industries Board remained headless and impotent. Then in March 1918, the president detached it from the Council of National Defense, placed it directly under himself, and selected Bernard M. Baruch as its chairman.

It was a brilliant choice. This tall southern gentleman with his dark handsome looks, his proud bearing and courtly manners, and a great fund of tales that he recounted in his deep cultivated voice had a gift for entertaining and cajoling people into accomplishing difficult tasks. Deferential to those above him, loyal to those who worked with him and for him, he was blessed with a knack for sincere flattery. Baruch could persuade a busy man to perform a task for him by explaining how significant the job was and by letting the hearer know "in confidence" how important he was to him. A master of the indirect approach, the WIB chairman found ways to inspire others to suggest changes he himself wished to have made. "It is a great thing," he said, "to always let a man think he has a new idea, show him how to do it, and pat him on the back and tell him how well he has done it." Baruch hired able younger men who knew a good deal about their own industries and allowed them considerable freedom to do the agency's work. He did his best to give them a sense that they were wanted. He built esprit de corps.

Baruch was rich. He had made a fortune speculating in minerals and in other astute investments based on thorough, sometimes inside knowledge gained through a large network of connections, and he used his wealth to expand and solidify that network and to reward associates and employees and those from whom he drew his strength. Sometimes the reward was a gift of ducks from his South Carolina estate or the opportunity to dine with him or share his cab or his private railway car or to socialize in his luxurious suite. When appropriate, a generous gift of cash might be forthcoming, like the $12,500 he gave to support Wilson's candidacy in 1912 or the $35,000 or so he donated to the president's 1916 campaign. Wilson liked him, and Baruch used his relationship to the president to strengthen his own authority and to accomplish important public purposes.

Baruch had other skills and qualities invaluable for running the War Industries Board. He was a resourceful bargainer. The wide knowledge he had acquired in his business and as a commissioner for raw materials for the National Defense Advisory Commission had given him detailed understanding of much of American industry. While other businessmen respected his wealth and liked him personally, as a Wall Street speculator, as an independent operator, and as a Jew, he could never be at the center of the American business establishment. In the circumstances, that was an advantage, for he needed to be thought of as detached from the great corporations he helped to regulate. Baruch had an excellent grasp of the value of appearances. By selling his holdings and giving part of the proceeds to charity and by admitting to Congress that he had been a speculator by trade, he allayed criticism. A master of public relations, he hired people to help him persuade others of his importance and disinterestedness and of the significance and power of the War Industries Board. He was also adept at deflecting unpleasant questions about the authority President Wilson had given him and his organization.

When the president appointed Baruch chairman of the War Industries Board, he told him to find out what other federal agencies needed to fight the war, to represent the government supply departments in purchasing, and to oversee the placement of contracts. Baruch was also charged with locating scarce materials and assuring rapid delivery of orders. Yet Wilson had no intention of presenting Baruch, or any subordinate, with absolute authority. Crucial decisions still required approval from the White House.

Wilson told the War Industries Board chairman to interfere as little as

possible with existing operations of the War Department and other purchasing agencies. Baruch was not to pursue policies that a large number of
the other WIB members opposed. Even after Congress passed the Over-
man Act, which appeared to accord the president the power he had already delegated to Baruch, the WIB chairman's authority remained subject to challenge.

Still, Baruch was in a stronger position than his predecessors, and
under his direction the War Industries Board extended its controls. Building on the work of earlier agencies, it compiled data about the current and
potential abilities of American plants to produce war materials so that by
mid-1918 it had classified twenty-eight thousand factories according to the
production processes for which each was equipped. It gathered estimates
of the kinds and quantities of goods the military services, the Allies, and
other consumers needed and tried to make sure that supplies were distributed to one group without crippling the others. The WIB negotiated
war contracts, helped plants convert to military production, arranged to
have new factories built, provided war producers with technical know-
how, and handled orders from the Allies. A conservation division helped
eliminate wasteful practices, particularly by encouraging companies to
make standard products. It arranged, for example, to cut buggy wheel
types from 232 to 4 and buggy body styles from 20 to 1. When the war
ended, the WIB was preparing shoe manufacturers to turn out only three
styles apiece for men and women.

Gradually, the War Department became more cooperative. Even before Baruch's appointment, the army had begun to reform its supply
operations, transferring them from the bureaus to a new Division of Purchase, Storage, and Traffic. Developing more accurate methods of pre-
dicting the army's needs for matériel, the War Department passed these
predictions on to the War Industries Board. It also devised standard contract forms and new contracting procedures and sent as its representative
to WIB meetings General Hugh Johnson, who, as time passed, began to
get on well with civilian executives.

To perform the detailed work of matching suppliers with consumers,
the War Industries Board replaced the CND cooperative committees with
fifty-seven new committees arranged by commodities. Staffed with representatives of the War and Navy departments and other purchasers and
with dollar-a-year men, the Commodity Committees negotiated on behalf
of the United States government with War Service Committees—actually
trade associations—which represented suppliers. It was easiest to make
agreements and enforce them for industries dominated by a few companies whose leaders had fixed prices and allocated production before the
war. When industries were competitive or disorganized, the WIB encour-

aged companies to form trade associations and negotiate collectively with the board's representatives.

Together the Commodity Committees and War Service Committees discussed trade secrets, costs, and other information that would enable companies to cooperate in war production. They worked out among themselves who would receive contracts, the price the government would pay, and priorities for delivery and then passed their recommendations to higher-level WIB divisions and to a separate price-fixing committee that included Bernard Baruch. The recommendations of the lower committees were almost always approved. The reformed system still invited collusion, and in peacetime the participants might have faced prosecution under the Sherman and Clayton acts. But with the War Industries Board overseeing their actions, competition gave way to cooperation, and, as in the case of the Railroad Administration, enforcement of the antitrust laws was effectively suspended.

4 ‖ The War Economy: Motivations and Results

By the summer of 1918, the structure for managing the war economy was essentially complete. The WIB and other agencies were able to supervise decisions of private industry; to regulate consumption of fuel, agricultural products, and other materials vital to war; to stimulate military production; to manipulate commerce and finance; and to control the nation's transportation system. These actions constituted a substantial shift, a movement toward more centralized management and more government control than most American businessmen would have accepted before 1917. At the same time, the change to a war economy created great difficulties for many American companies, since the war agencies expected them to relinquish civilian customers on whom their long-term prosperity depended, to abandon distribution networks developed over many years, to produce items they had never made before and which they could not hope to sell in peacetime, and to do business with a government that was likely to cut off its orders when the war ended. Under these circumstances, it was natural that some companies hesitated to convert to war production. But the significant fact is that so many American businesspeople not only accepted the wartime economic system but also, in many cases, eagerly took part in it. What impelled them to do so?

One answer is patriotism. Grosvenor Clarkson, a public-relations man who served as secretary and later director of the Council of National Defense, stressed this motive in his 1923 history, *Industrial America in the World War: The Strategy Behind the Lines*. Industry, he wrote,

> strove for its own greater self—the Nation. . . . Subjected to the test of patriotic service, the most sordid business men, even the branded ghouls of the under-world of business, did their duty rather than face the contempt of the trade. . . .

Clarkson recalled that a complainant who came to the War Industries Board bursting with indignation would cheerfully reverse himself when the public-welfare point of view was presented. "The spirit of service for the common good was ultimately supreme in all men in those times," Clarkson remembered. "The American business man never showed himself more favorably than in his relations with the War Industries Board. When the hour of sacrifice came, he gave his business to the Government as freely as he gave his sons."

There was considerable substance to these statements. Many businesspeople felt patriotic pride in helping to conduct the greatest war within memory. Food Administration director Herbert Hoover informed the president that in agriculture "key men of the foreman, manager and ownership type," whose departure would interfere with food production, were too patriotic to apply for exemption from the draft. He also told Wilson that some hundred fifty grain dealers and grain-elevator operators had approved wholeheartedly a Food Administration program for control of the wheat and rye businesses of the United States, even though that program would put some of them out of business. Wilson commented: "It certainly was a very unusual and stimulating exhibition given by the grain dealers and elevator men. . . ." The records of military and civilian volunteer service by businesspeople from communities all over the nation provide ample testimony to their patriotic support of the war. Yet patriotism does not entirely explain their acceptance of the wartime economic system.

One problem with the patriotism theory is that some businesspeople acted as if business came first and the war second or not at all. Clarkson hinted at this when he said the spirit of service for the common good was *ultimately* supreme. He made the point explicit by referring to "swindling jobbers of lumber," "gangs of bloodsuckers," and businessmen who "regarded the war as a golden opportunity for filling their coffers." To President Wilson, the willingness of grain dealers and grain-elevator men to sacrifice their business was "very unusual." He was in a position to know. A Federal Trade Commission member had recently told him that speculators were keeping thousands of railroad cars filled with coal off the market, creating an artificial coal shortage to force up the price. Through a tip from a French official, Bernard Baruch discovered that when the United States was critically short of oceangoing vessels, about two hundred thousand tons of American shipping were being withheld from war service for larger profits in civilian trade. In the knit-goods industry, several manufacturers refused government contracts because civilian business was more lucrative. Acid manufacturers threatened not to sell to the government unless it offered a "stimulating" price. Attorney General Gregory urged the president to continue prosecuting antitrust cases by pointing to "price extortions which have been practiced during the present emergency. . . ." Even Herbert Hoover, who believed he had the support of people in the food trade, remarked, eight months after the country had entered the war, that "profiteering still runs rampant in other branches of commerce" and that "the law of supply and demand has been replaced by the law of selfishness."

For certain businessmen, feelings of opposition to government labor policies proved stronger than patriotism. Even if the government paid them to do it, some companies would not make changes in working conditions or allow their workers to organize. Southern cotton mills declined army contracts that limited child labor. Some manufacturers said they would refuse government war contracts that required them to provide an eight-hour day. Sir Stevenson Kent, a British coal-mine and steamship operator, wrote Secretary of War Baker in November 1917, after a trip through the Midwest, during which he had spoken with labor-union members and chambers of commerce, that "we have found almost without exception a firm determination amongst the employers to maintain the conditions of the open shop"* and that, in general, "employers in this country attach so much importance to this fact that the winning of the war is almost a secondary consideration to them."

Several years later, Baker recalled that during the World War "the 'public interest' never got any sympathy from employer or employee groups . . . ," a view also expressed by the head of the United States Shipping Board, Edward N. Hurley, a former manufacturer and president of a manufacturers association. "I find no patriotism on either side when it comes to money," Hurley remarked. Employers he encountered did not want to cooperate with the government. They insisted that it pay all their bills, including the expense of higher wages. "After my work in behalf of the employers of the country, to find this feeling prevailing in most quarters is . . . disheartening."

Other business leaders shared Hurley's view. The president of the National Lumber Manufacturers' Association, Robert H. Downman, who headed the Council of National Defense lumber committee, wrote the president of the California Redwood Association:

> I feel quite sure that if you will stay in Washington for a little time and become active on this committee you will realize what a hard job we have had and about come to the conclusion that patriotism, in most cases, is nothing much more than a thin veneer.

*A shop where union membership is not required for employment; or a shop where union members are not knowingly employed.

Even Bernard Baruch remarked that the stories of commodity groups enthusiastically accepting government prices were myths and that his own recollection was of long and tedious bickering. Patriotism, then, does not sufficiently explain why so many businessmen participated. One has to note additional motivating factors: threats, coercion, and particularly, incentives.

Economic managers wielded several coercive instruments, such as the power Congress gave the president to seize property for military purposes. As we have seen, in the early phase of mobilization the armed forces used this power frequently, making hundreds of compulsory requisitions, sometimes appropriating whole plants and hiring special companies to run them. Other agencies did the same; for example, the Shipping Board, which commandeered a substantial part of the American merchant fleet. Once, when a Smith & Wesson arms plant refused to abide by a judgment of the National War Labor Board, the army was dispatched to take it over.

The War Industries Board frowned on such practices, and as the board's influence increased, the government used the power to commandeer sparingly. Board members claimed it was illegal to take property for anything but narrowly defined military aims. They argued that commandeering simply could not work as a way of running decentralized industries that contained thousands of small firms and was unsuitable even when the business was large and centralized. Baruch remembered that when the board considered seizing a large industrial plant, it rejected the idea after posing these questions: Who will run it? Do you know another manufacturer fit to take over its administration? Would you replace a proved expert manager by a problematical mediocrity? After you have taken it over and installed your government employee as manager, what greater control would you have then than now?

At the same time, the War Industries Board and other war agencies were perfectly willing to threaten commandeering. The WIB passed a resolution to have the steel plants taken over if their owners refused to charge the prices it recommended. In 1918, when an auto manufacturer refused to cut back steel consumption for civilian pleasure cars, the board hinted it might seize his raw materials.

Another reason why businessmen cooperated with the board and other government agencies was to avoid bad publicity. At the beginning of the war, many Americans regarded large corporations with suspicion and fear, as corrupters of governments and plunderers of consumers and small businessmen. War work in hundreds of voluntary agencies, together with propaganda extolling their wartime service and lauding their patriotism, offered a way to overcome this legacy of the prewar years.

While some of this propaganda emanated from business groups themselves, the federal government issued much of it. But a government that created good publicity could also damage reputations, a point the administration and its agents repeatedly used when bargaining. The War Industries Board warned a lumber manufacturer that if he did not improve his attitude, the government would inflame his hometown against him. The president publicly denounced uncooperative businessmen, declaring that those who fixed ocean shipping rates too high had taken "the most effective means in their power to defeat the armies engaged against Germany," and warned recalcitrant manufacturers that "those who do not respond in the spirit of those who have gone to give their lives for us on bloody fields far away, may safely be left to be dealt with by public opinion and the law. . . ."

The war bureaucrats threatened bad publicity more than they used it. While they were demanding unity and sacrifice from soldiers and workers and the rest of the American people, and while socialists and other radicals were charging that working people were fighting a war for the rich, it was awkward to denounce leading companies. Government agencies, with their ties to corporations, preferred more subtle, indirect ways (like the manipulation of priorities and the channeling of capital) of influencing business decisions.

Priority control was one of their favored instruments. By determining the priority—or order—in which items were to be delivered, war agencies were able to pool, ration, open bottlenecks, halt competition that interfered with the war effort, restrain hoarding, conserve fuel, curtail production of nonessential materials, and make shipping space available for military supplies. Priorities helped the government disperse war plants, regulate transportation, and control prices throughout the economy.

War agencies gradually extended government control of priorities until, by the summer of 1918, they had covered virtually the entire American economy. Their ability to do this derived in some cases from legislation (such as the Lever Act, which empowered the Fuel Administration to determine who received fuel supplies; and a military appropriations act, which made it possible for the Railroad Administration to allocate shipping). In other cases, the War Industries Board arranged with major suppliers, like the steel industry, to let it decide how producers of those supplies would distribute products to their customers. It used its ability to control steel orders to curtail shipments to such other industries as those producing passenger automobiles, vacuum cleaners, corsets, ice-cream freezers, and talking machines. By September 1918, the War Industries Board had secured the right to coordinate commandeering, so that in a case where a company refused to comply with its priority system, the WIB could have it seized.

Once the priority systems were in place, the government actually did not have to commandeer or even to criticize companies publicly to make them do what it wanted. Instead, it could offer them scarce supplies if they cooperated and if they refused, could, at least in theory, deprive them of fuel and power, cut off their raw materials, make sure no freight cars came near their sidings, and divert business to their competitors. It could say, in effect, "Produce what you want. We won't make you manufacture war materials. But if you don't, we will make it hard for you to continue your business." It could induce volunteering.

Yet the government's bargaining position was far from all-powerful. Even if it had wanted to, the War Industries Board could not have run a totally centralized system, for it lacked enough information about the nation's businesses, it was swamped with paperwork, and its legal authority to set priorities for certain goods was questionable. The president never granted the board power to override the wishes of the armed services. It could not compel the largest American companies to obey its edicts because if they chose not to cooperate, they could make the whole war effort collapse. In 1918 the WIB abandoned attempts to set priorities case by case and instead sent out an official priorities list, rating seven thousand plants in order of their estimated importance to the war effort. It allowed manufacturers to decide what priorities their customers' orders would receive and permitted suppliers to determine which manufacturers' orders to fill. Leading companies and trade associations, which had far better intelligence about their industries than the WIB did, decided who got what.

Even when an industry declined to convert to war production, the War Industries Board was extremely reluctant to have it shut down or taken over. It much preferred to negotiate, using its ability to withhold priorities and its power to recommend commandeering as bargaining tools. A bitter controversy between the automobile industry and the board illustrates this kind of negotiation.

By the summer of 1918, no more than 10 percent of the nation's auto companies had converted to war work. The manufacturers continued to produce civilian passenger cars ("pleasure cars" as they were called at the time), consuming large quantities of materials that might have been used for war production and employing skilled workers who could have been turning out ships and weapons. Some auto companies were stockpiling steel. From the point of view of many carmakers, this made perfect sense. One of their largest competitors, the Ford Motor Company, was receiving steel for war contracts while continuing to produce pleasure cars. They believed that if they retooled to produce military vehicles, their civilian distribution systems would disintegrate, leaving them when the war ended with an ample supply of army trucks that no one wanted while Ford was selling passenger cars through its intact dealer network.

The auto companies offered to cut pleasure-car production to half of their 1917 output. The War Industries Board told them to convert entirely or it would limit their coal shipments and have the steel companies (which had given the board control of priorities) reduce their supplies of steel. This infuriated industry representatives. They argued, correctly as it turned out, that there would be plenty of steel to make cars and military goods if the steel industry produced what it was capable of producing and if the government itself stopped hoarding raw materials and unclogged the nation's transportation system. Hugh Chalmers of the National Automobile Chamber of Commerce told the War Industries Board, "That is the beginning of the end. . . . It is absolutely confiscation of the industry. . . . You might as well appoint a receiver for the state of Michigan, . . . [W]e are not going to sit here and bow our heads. We are the third largest industry in this country, and with all the men we employ and all the obligations we have to our parts people, all the money we owe our banks we can not surrender to this." Chalmers added that many steel manufacturers were disposed to look out for their old customers who would be buying steel after the war and would ship to them regardless of the WIB's priorities.

After months of haggling, the board and the auto companies compromised. The WIB promised to allow them enough steel and other materials to clear out their existing stocks. In return, the companies, weakened by shortages of all kinds of raw materials, agreed to cut pleasure-car production to a quarter of the 1917 level, to conserve as much steel as they could, to release excess raw materials to needy industries, and to submit reports to the board under oath.

In theory, the WIB could have wiped out nonessential firms or forced whole industries like the automobile industry to abandon civilian production completely. But this would have made a larger change in the nation's economic structure than business leaders, including those who were running the war economy, were willing to accept. The war managers worked out arrangements that enabled companies at the end of the conflict to regain something like their prewar share of the market. Consequently, not only the auto industry but even industries that produced no war products at all received in exchange for some concession (for example, an agreement to conserve scarce materials,) at lease the trickle of goods and services they needed to survive. Those companies that wanted to do more than just survive converted to war production and were moved toward the

top of priority lists. Thus, the War Industries Board secured the cooperation of businessmen through what one of them called "the involuntary
voluntary method."

The ability to influence the flow of capital gave wartime government
managers another means for moving businesses in desired directions without coercing them directly. A Capital Issues Committee arranged with
investment houses to pass upon their proposed issues of stocks and bonds.
Though the committee had no enforcement powers, most of the companies it dealt with abided by its recommendations. A War Finance Corporation arranged private loans for businesses the government considered
essential and in some cases lent government money directly. The Emergency Fleet Corporation, the War Department, and other government
purchasing agencies eased borrowing problems of war contractors by letting them charge costs directly to the Treasury, paying them all or part of
their fees in advance, or even constructing factories for them. In these
ways, while attempting to keep capital away from nonessential business,
the government directed funds toward firms producing for war.

The makeup of the wartime regulatory agencies provided the business
community with a whole series of incentives to cooperate with the government. The groups that bargained on the government's behalf consisted
mostly of dollar-a-year men from the higher ranks of American corporations. The chairman of the Council of National Defense aluminum committee, for instance, was the president of the Aluminum Company of
America; the chief of the War Industries Board agricultural-implements
section was a former manager of Deere and Company, a leading producer
of agricultural implements; the head of the WIB rubber section had for-
merly headed the Fisk Rubber Company. These executives negotiated as
agents of the United States with people from their own industries, even
from their own companies. Executives from Lackawanna Steel, Cambria
Steel, and Youngstown Sheet and Tube sat on a WIB steel committee that
negotiated with a trade association that included the presidents of Lackawanna Steel, Cambria Steel, and Youngstown Sheet and Tube.

Such arrangements proved advantageous to the companies and the
dollar-a-year men alike. The latter found opportunities in government
service to develop more intimate knowledge about their industries, including trade secrets of their competitors, while making contact with potential
future employers. The executives they negotiated with were reassured by
the fact that the persons who sat opposite them at the bargaining table
were practical businesspeople, not theorists or civil servants, people who
understood the industries they regulated, sympathized with those industries, and were not inclined to overhaul the private enterprise system any
more than was required. The companies also recognized that by sending

their executives to serve on the WIB, they could gather intelligence about the government's plans and influence government actions; and they understood that if they failed to do so and let other firms provide all the dollar-a-year men, their competitors would determine policies for the entire industry.

The fact that these regulators served two masters disturbed people who feared what business self-regulation would do to farmers, small businessmen outside the system, and consumers. To appease those who worried about the enduring problem of conflict of interest, the order that established the War Industries Board instructed the board to eliminate direct or indirect connections between the commodity committee members and the industries with which they dealt. Still, ways were found around the prohibition. In addition, by pretending that the war service committees were simply advising it rather than playing a key part in negotiating contracts, the War Industries Board bypassed the provision of the Lever Act that prohibited conflicts of interest. In June 1918 the board issued a resolution requiring its members to declare their interest in transactions they were negotiating, but no one made a serious effort to enforce it. It could not be enforced because the system for managing the war economy depended on the technical knowledge, managerial skills, and willingness to cooperate of people and companies for whom conflicts of interest were inevitable. Under these circumstances, companies that dealt with the government were able to make very substantial returns. For American businessmen, profit was one of the chief inducements to aid their nation in the war.

Every company that sold to the government did not become rich overnight; but certain types of businesses did especially well: those holding scarce essential commodities; low-cost producers in industries where the war agencies fixed prices so high that even inefficient companies could make a profit; companies whose products or services were in great demand but whose costs were difficult to estimate; and firms working under "cost-plus" contracts that allotted profits as a percentage of costs, thus requiring the government to pay them more for profits as their expenses rose.

Gains sometimes were small but elegantly achieved, as in the case of a company that used virgin lumber to cremate dead horses under a cost-plus contract. Often the rewards were very large. After-tax profits of forty lumber companies averaged 17 percent on investment in 1917, while the earnings of twenty-one copper companies rose from an average of 12 percent in 1913 to 24 percent in 1917, despite a prolonged copper strike. The excess-profits tax was supposed to siphon off some of the more lavish returns, but the government allowed war contractors to hide profits as untaxable costs; it permitted firms to limit taxation by investing in certain types of Liberty Bonds; and it negotiated reductions of wartime taxes after the Armistice. These profits were only part of the gains the government allowed, since it also paid the costs of developing know-how and provided companies with machinery and buildings that could be used in peacetime. To the chemical industry, it awarded the prize of German patents.

The steel-industry case of 1917 shows how generous the government could be. The price of steel had begun to increase before the United States entered the war and rose spectacularly afterwards, far outpacing the cost of production in certain mills. Since steel was basic to the war economy, the government had to bring its price under control, but the steel managers proved difficult to bargain with. Not only did they resist lowering their charges to the armed forces and the Emergency Fleet Corporation, but they also insisted on selling to other customers at even higher prices than they charged the United States government.

For weeks, the administration made a display of angry concern. Baruch and the president directed ominous threats at the industry while Senator Atlee Pomerene of Ohio introduced a resolution to allow a takeover of steel plants that levied excessive charges. At the request of the secretary of the navy and the Shipping Board, the Federal Trade Commission investigated steel-industry costs, and the chairman of the Shipping Board threatened to commandeer steel mills if they did not drop their prices to a level suggested by what the FTC found. This preliminary fencing ended in September when steel executives met with the price-fixing committee; negotiated prices on basic steel, pig iron, iron ore, and coke; and agreed to charge all other customers what it charged the federal government.

On the surface, the settlement was a victory for the War Industries Board and the public. In fact, it was a victory for the industry. The companies lowered prices drastically, reducing steel plates, for instance, from $11.00 per hundred weight to $3.25. However, at the time it agreed to the reductions, the War Industries Board knew from the report of the Federal Trade Commission that U.S. Steel could have made a nearly 50 percent profit by selling plates at $2.90.

The WIB thus gave government approval to prices that guaranteed extraordinary earnings. Just how large those profits were can be judged from industry records for 1918, after government prices were set. The average return on investment that year for steel manufacturers was 20.1 percent. U.S. Steel paid a 14 percent dividend on common stock. Four officers of Bethlehem Steel divided a bonus of $2,100,000.

Industry justified these returns with the argument, accepted by the War Industries Board, that prices had to be set in a way to maximize production. Since it was impossible to go into every mill in the country, find out what its costs were, and pay it a price that would give it the same rate of return as every other company in the industry, the WIB decided that a single price had to be set for each steel product, high enough to make small-scale operations pay, even if this meant large profits for bigger, more efficient producers. The difficulty with this argument is that small producers not only made a profit at government-set prices, but some of them made a far larger profit in proportion to investment than any of the big steel companies. Congressional investigators discovered after the war that the return on investment for one group of ten small mills ranged from 30 to 319 percent. It was this kind of gain that led an executive of the McKinney Steel Company to say, "We are all making more money out of this war than the average human being ought to."

While large guaranteed profits induced American businessmen to support the war, some of their leaders saw in the wartime-managed economy a way toward a kind of gain that would continue long after the guns stopped firing. To these men, advocates of what might be called a New Capitalism, participation in the war effort provided the chance to create a better form of business enterprise.

Proponents of the New Capitalism wanted to move beyond the insecurity and unpredictability of nineteenth-century business practices toward a stabler and more efficient system. For earlier haphazard methods of organization and production, they wished to substitute "scientific management." To help manage workers in a more rational manner while avoiding the violent labor disputes that had plagued American industry before the war, they instituted such labor policies as stock-sharing and bonus plans designed to make workers identify with their companies, and they cultivated union leaders who rejected class warfare and socialism. New Capitalists wanted to do away with the conflict in the marketplace that periodically drove down prices, made it hard to plan production, and forced the economy to swing between crashes and booms. They promoted trade associations that were intended to share technical information, to lower costs by standardizing products and production methods, to stabilize markets, to control price fluctuations (especially downward fluctuations), and to ensure steady returns year after year. Through efficiencies created in combined activities, New Capitalists hoped to win a larger share of foreign markets. There would still be competition, but it would be based not on price but on advertised differences between the essentially similar items that each industry would produce.

To enforce cooperation, to avoid regulation by dozens of states with their varying rules, and to pacify a public that in the prewar years had become wary of collusion and price fixing, New Capitalists invited the federal government into their system. They believed they could prevail upon it to allow business groups, under its friendly oversight, to run the economy while regulating themselves. Federal supervision offered a way around the Sherman and Clayton antitrust acts, for it would enable executives to take part in collective actions that would otherwise expose them to fines, prison terms, and damage suits. The public was expected to feel that the government would be protecting their interests.

To New Capitalists, war offered a chance to build what a United States Chamber of Commerce official described as an entirely new relationship between government and industry. Howard E. Coffin, a founder of the Council of National Defense and a leading advocate of the New Capitalism, declared shortly before the council held its first meeting: "[I]t is our hope that we may lay the foundation for that closely knit structure, industrial, civil and military, which every thinking American has come to realize is vital to the future life of this country, in peace and in commerce, no less than in possible war." The history of the Railroad Administration and the War Industries Board shows how, during World War I, Coffin's dream began to materialize. As the president of Republic Steel observed after President Wilson announced the 1917 steel price agreement, "[T]he principle of cooperative regulation has been established with Government approval."

Thanks to the Wilson administration's policies, the Great War became what Grosvenor Clarkson called "a golden age of harmony" when "government prices made a living possible for all except the submerged tenth of shoestring industry; and executives, relieved from the nightmare of menacing losses, were free to give their attention to quality and quantity of product." This was a time, Clarkson recalled, "when men who dealt with the industries of a nation . . . meditated with a sort of intellectual contempt on the huge hit-and-miss confusion of peacetime industry, with its perpetual cycle of surfeit and dearth." The Great War had produced a welfare state for the American business community.

The chance to build, run, and profit from this war welfare state was a major incentive for many business leaders to cooperate in the managed economy. But one must ask why the Wilson administration helped bring it into being. When Woodrow Wilson first ran for the presidency in 1912, among the tenets of his New Freedom program had been encouragement of competition and keeping government free from business control. Some of his opponents, including Theodore Roosevelt, had argued that industrial combination was a necessary and permanent development and that the way to handle the collectivization of American business was through government regulation. To this, Wilson had replied: "If the government is to tell big business men how to run their business, then don't you see that big business men have to get closer to the government even than they are

now? Don't you see that they must capture the government, in order not
to be restrained too much by it?"

As the country moved toward war, President Wilson continued to
express concern about business domination of government. His secretary
of the navy recalled him as saying:

> If we enter this war the great interests which control steel, oil, shipping,
> munitions factories, mines, will of necessity become dominant factors, and
> when the war is over our government will be in their hands. We have been
> trying, and succeeding to a large extent, to unhorse government by privi-
> lege. If we go into this great war all we have gained will be lost, and neither
> you nor I will live long enough to see our country wrested from the control
> of monopoly.

Yet such concerns did not stop Wilson from welcoming dollar-a-year men
into the advisory commission of the Council of National Defense or from
allowing representatives of big business to dominate the War Industries
Board. How do we reconcile the president's words in 1912 and Secretary
Daniels's memory of his prewar views with the actions of his war
administration?

Woodrow Wilson was a highly realistic, pragmatic man, who accepted
the basic structure of economic power in the United States. He had never
intended, not even in 1912, to transform American capitalism into a system of small competing companies or to make war against large corporations. He supported efforts to clarify the antitrust law and to forbid certain
practices (like interlocking directorates) that impeded competition. During the prewar years, he pursued antitrust suits against major companies.
He listened to and cooperated with those in his party who distrusted
corporate power, people like Daniels and the agrarian radicals who fol
lowed William Jennings Bryan. Wilson paid attention, when it was practical to do so, to labor unions and social reformers. But in both his terms of
office, he also sought cooperation from business leaders. He supported the
establishment of a Federal Trade Commission, which some corporate
leaders hoped to use as a source of guidance and intelligence, and appointed conservative businessmen to run it. The Federal Reserve system, one of
the major achievements of his first administration, strengthened banks and
assisted other business interests. Wilson also endorsed trade associations
and encouraged American companies engaged in international trade to
cooperate with one another.

Once the United States was in the war, he recognized that his administration could not continue prewar efforts to promote competition, because military success depended on the support of large companies. At-

torney General Gregory recommended in the fall of 1917 that the administration continue antitrust suits against coal, steel, and other companies that were seizing control of the necessities of life and driving up prices. The War Industries Board declined to support these prosecutions, and the president, realizing that to vindicate the antitrust law against major American corporations might disorganize economic mobilization even further, decided, as Gregory recalled it, to put antitrust suits to sleep until the fighting ended so that these giant companies would have no excuse not to cooperate in the war. When Congress was considering the Lever bill, Wilson sought to dissuade it from prohibiting conflict of interest on the ground that dollar-a-year men who advised the Council of National Defense might leave the government. That, Wilson claimed, would be a "calamity," seriously, perhaps fatally embarrassing to the government, which had to have the cooperation of the men who were in "actual control of the great business enterprises of the country." Verifying his 1912 suggestion that the regulated would control the regulators, he wrote Senator Kenneth McKellar of Tennessee in July 1917 that "we are dependent" on the advice and assistance of "the chief businessmen of the country. . . ."

Wilson did not enjoy conceding so much to corporate leaders. He spoke out against profiteers. Yet, his outbursts against wealth and power were sporadic and mostly oblique. In December 1917, after eight months of bitter struggle to secure the cooperation of the automobile and steel companies and other producers and after hearing repeated reports about the greed of speculators, the president gave a speech at the Gridiron Club before politicians and leaders of industry and finance. He talked of the need for sacrifice and criticized selfishness. "The thing that buoys me up every day," he said, "is the consciousness that the people of the United States would lose heart in this war if they thought we were fighting for anything selfish at all."

He related to this audience a dream he had had about one of the "prima donnas" he was forced to pacify to get the business of the government done. He explained that in this dream he had decided to publish "a list of gentlemen who preferred their private sensibilities to their public duties. . . ." The prima donnas were by no means only businessmen. They were congressmen and executive-branch officials, union leaders, American generals, and representatives of the Allies. But among the chief figures the president had to cajole were men from the business community on whom he knew he utterly depended because of their knowledge, their connections, and their ability to produce war supplies.

This dependency forced the administration to make large concessions in the bargaining it carried on with American companies. Unable to expose the control that corporations exercised in the war economy without losing their assistance and encouraging socialists and other dissenters, unwilling to provoke a debate in Congress and the country about how much control war producers ought to have, the president tried to limit corporations' power in indirect and negative ways. Even after the crisis of the winter of 1917–1918, he made no effort to transfer control over military procurement to the War Industries Board, and he never asked Congress to give the board the kind of legal authority the armed services enjoyed. Thus, he allowed the armed forces to check the power of the business-dominated war agencies. Even if it entailed some inefficiency and permitted conflict of interest, he wanted a temporary structure for industrial mobilization run by volunteers who would go back to their companies when American industry had turned out what was needed to win the war.

How effectively did the wartime system for managing the economy contribute to victory over the Central Powers? For understandable reasons, the answer is mixed. From 1916 through 1918, the gross national product of the United States rose by less than 1 percent; total farm output increased slightly; and the output of copper smelters, pig iron, iron ore, and rolled iron and steel fell. Although Emergency Fleet Corporation yards produced nearly three million tons of new shipping a day in the last half of 1918, not one ship from the huge Hog Island shipyard was ready for service when the shooting ended. The country spent some seven billion dollars for ordnance; its arsenals turned out five thousand modified British Enfield rifles a day; yet the American Expeditionary Force used French artillery and projectiles most of the time. Despite the expenditure of many millions of dollars, aircraft manufacturers produced only two thousand units in 1917 and fourteen thousand in 1918, far below administration projections.

Commenting on a shortage of tanks, AEF commander General John J. Pershing remarked: "It seems strange, . . . with American genius for manufacturing from iron and steel, [that] we should find ourselves after a year and a half of war almost completely without those mechanical contrivances which had exercised such a great influence on the western front in reducing infantry losses." British prime minister David Lloyd George observed that "one of the inexplicable paradoxes of history" was that "the greatest machine-producing nation on earth failed to turn out the mechanism of war after 18 months of sweating and toiling and hustling. . . ."

Still, considering the circumstances, the United States did reasonably well. Instead of arguing that American industry produced too little before the Armistice, one might say that the Armistice came too soon for industry to reach peak production. A nation whose economy had been operating at near-capacity before April 6, 1917, needed months, if not years, to arrange contracts, build or convert plants, construct machine tools, accelerate munitions production, adapt weapons to changes demanded by people at the front, and deliver the instruments of war. To have armed and supplied all its troops completely in the spring of 1917, the United States would have required, years earlier, a vast military-industrial network that isolationist, antimilitarist taxpayers would not have tolerated and an economy that produced less for civilians. The failure to turn out more equipment in wartime, though troublesome, was understandable. It was far from disastrous because Allied factories, sustained by American dollars and supplied with American raw materials, were turning out enough armaments to equip many of the doughboys who poured into France.

Some of the men who ran the wartime economic system wished to keep parts of it going after the Armistice. Secretary McAdoo was so pleased with the Railroad Administration that he recommended continued government operation for five more years. A number of businessmen connected with the WIB suggested using war agencies to maintain prices during reconversion. Some of the New Capitalists did not want to go back to the prewar competitive system, at least not without drastic changes. George Peek of the WIB hoped Congress would permit industrial cooperation (which meant altering antitrust laws) so that lessons learned during the war "may be capitalized in the interest of business and the public in peace times." Peek wanted to continue government-supervised price fixing for certain products. Heads of War Industries Board sections discussed with one another plans for increasing overseas trade, revising antitrust legislation, and arranging for the government to demobilize workers so that they would fit most efficiently into the postwar economy. Two days after the Armistice, representatives of the steel industry came to Washington to ask the board to stimulate government buying, maintain price agreements, and take any other steps necessary to provide an orderly transition to a peacetime economy.

On these issues, the business community was split. Alongside advocates of the New Capitalism were other, more traditionally minded businessmen who, now that the government was no longer in a position to hand out lavish contracts, wanted to free themselves as quickly as possible from federal regulation. Some auto manufacturers wanted to reconvert completely to pleasure-car production before their competitors took over the postwar market. Shippers preferred to send their goods along routes and at times that they, not the Railroad Administration, chose, while railroad owners demanded their properties back. Industrialists who felt they could raise prices rapidly in postwar markets wanted pricing restraints removed. Companies wished to deal with their workers without further intervention by government.

Most dollar-a-year men wanted to return to their prewar jobs, and Baruch and some of the section chiefs of the War Industries Board realized that, without the impetus of war, any regulatory system that continued after the Armistice would have to be staffed with government bureaucrats. Such people might be less favorable toward industry than WIB volunteers had been. Finally, there was the danger, noted by Bernard Baruch, that if the agency he headed did not close down immediately, the public would discover how corporations had used it, thus tarnishing the images built up during the war of selfless patriotic corporations and a disinterested War Industries Board.

Wilson told Baruch that the sooner business and government separated the better and ordered the WIB to end its activities by the beginning of 1919. But to ease reconstruction problems somewhat, he established a National Industrial Board, intended to continue some WIB functions temporarily. The president abolished the Food Administration and informed Congress, late in 1919, that if it did not do something about the railroads, he would return them to private control by March 1, 1920. Congress then passed the Transportation Act of 1920 which restored the roads to private management while authorizing a stringent program of federal railroad regulation. It also handed back the shipping industry to its former owners.

What, in sum, were the effects of the Great War on the American economic system? The immediate consequences were immense. War led to the near collapse of an unprepared economy and produced a new regulatory structure that repaired the damage and managed mobilization. The people who operated this structure, chiefly business executives assisted by academic and other experts, increased production of certain goods, decreased production of others, regulated consumption, and coordinated distribution. At times, the war bureaucrats used threats and other coercive tactics to accomplish their objectives, but in dealing with organizations of great power and with an economic system in which they had a stake, they usually followed an indirect approach, offering high priorities, ample profits, and the chance to develop a New Capitalism as incentives to cooperate. Potentially the most important result of all was a lesson deposited in the historical record: overseas wars can be beneficial to American business—for the profits they generate and for the security and stability a war welfare state affords to those in a position to take advantage of it.

5 ‖‖ The War and Social Reform: Workers and the Poor

At one time, historians believed World War I had delivered a coup de grace to the Progressive movement, the series of campaigns that had arisen before 1914 to reform American society. There is considerable evidence for this view. We have noted the way business groups, assailed by reformers in the prewar years, took command of the councils of defense, the War Industries Board, and other volunteer agencies and used them to reassert their influence. We have seen how people who wanted to make fundamental reforms in the relationships between social classes (for instance, members of the Industrial Workers of the World and certain elements of the Socialist Party) were persecuted during the war years. Yet when we look at what happened in those years to particular elements of the reform movement, we see that the effects of the Great War on progressivism varied greatly and, at least in the short run, were sometimes beneficial. This was the case for campaigns to aid working people and to do away with poverty.

In the quarter-century before the Great War, a movement spread through American society to mitigate the harsh effects of industrialization on workers and poor people and to disseminate its benefits more widely. Participants, who came from all strata and regions, included union leaders, upper-class humanitarians, socialists who wanted to change the nation's economic system, and people who feared labor violence and revolution. Just before the war, the social-justice reformers won a series of notable victories at every level of government, securing minimum-wage and maximum-hour laws, restrictions on child labor, workers' compensation, and special protections for female employees.

Mobilization threatened to nullify these achievements as employers tried to use the war to reduce labor standards, using the war as a reason. The National Association of Manufacturers asked Congress to authorize suspension of the eight-hour day on government contracts. The Council of

National Defense, dominated by business leaders, recommended that governors be allowed to modify protective restrictions in their labor laws and tried to prevent the states from trying to improve labor health and working conditions without its approval. In Massachusetts, businesses sent the state industrial board dozens of petitions to nullify restrictions on working hours and to suspend regulations protecting women and children, and there were similar requests in other states.

The state governments often looked favorably on these requests. Less than a week after the nation entered the war, the Vermont General Assembly passed a bill allowing the state labor commissioner to lift restrictions on working hours for women and children. A few days later the New Hampshire General Court authorized the governor to suspend state labor laws if the Council of National Defense requested it to do so. Massachusetts and Connecticut took similar actions. On the ground that war made it necessary for children to work on farms and in other enterprises, the California Legislature relaxed its requirements for compulsory education. Bills were introduced in New York State to suspend a railroad full-crew law, permit twelve year olds to stay out of school seven months a year to work on farms, and to restrain the state industrial commission from enforcing labor laws if enforcement hurt the war effort. An attempt was made to abrogate New York's fifty-four-hour law for women manufacturing military supplies. Though this legislation failed, the industrial commission exempted the Curtiss Aeroplane Company from legislation requiring a day of rest each week.

The federal government appeared to join the reaction against prewar labor reforms. In 1917 the United States Supreme Court upheld the right of companies to compel their workers not to join labor unions and in 1918 declared the federal child-labor law unconstitutional. Government agencies contracted with companies that paid low wages and worked their employees very long hours. The army bought uniforms made in sweatshops. The Bureau of Printing and Engraving paid its workers so little that they had to accept long hours of overtime to survive financially. Thousands of women spent thirteen to fifteen hours a day in the bureau turning out Liberty Bonds with no rest periods and only a few minutes for lunch. In an area where workers carried large sheets of paper from one machine to another, one young women told an investigator that she had to raise these sheets shoulder high six thousand times in a twelve-hour day.

If this were all that happened, the war would have been a disaster for the social-justice movement. Instead, the conflict proved extremely beneficial in certain respects: enhancing the arguments of reformers, enlarging the bargaining power of working people and their representatives, and bringing the benefits of a war welfare state to millions.

When business groups used the emergency as a reason for diminishing state protections for workers, reformers answered by talking about "efficiency," a word people regularly used to link their programs and interests to the war. For years, reformers had been saying that poverty made poor people inefficient by preventing them from living up to their potentialities. During the war, Florence Kelley of the National Consumers' League fought attempts to abrogate labor standards by claiming that poor working conditions hindered the production of war material. She pointed out how the British tried to increase munitions output by lengthening the working day until they discovered that after a certain number of hours, productivity of munitions workers declined. Then the British government set up an eight-hour work day for women in state factories, abolished the seven-day work week, and limited overtime, while improving working and living conditions of employees. The result was more efficient shell production.

This proposition carried considerable weight with American war bureaucrats, among them the army's chief of ordnance, who issued a general order to munitions makers that had been drafted by a scientific management expert and a consultant on the problems of women wage earners. This order, which imposed a series of protections for munitions workers, stated:

> Industrial history proves that reasonable hours, fair working conditions, and a proper wage scale are essential to high production. During the war every attempt should be made to conserve in every way all our achievements in the way of social betterment.

The Great War provided reformers and labor-union leaders with several other arguments: a war for democratic ideals had to be a war for industrial democracy at home in which employers would bargain with representatives of their workers rather than run the workplace autocratically; America must strengthen herself internally and set an example for the Allies by passing effective social legislation; lowering social and industrial standards would cost the nation more than military victory would gain; and America could not afford to weaken workers and other producers on whom the war effort depended.

More persuasive than any of these contentions were certain facts about the war. When the strength of armies depended, as much as it did in World War I, on social stability at home and on industrial production, no government could wage war effectively without the cooperation of most workers. The United States government ensured their support by allying itself with conservative labor unions connected to the American Federation of Labor.

The groundwork for this alliance had been laid years before when the AFL and its unions began to gravitate toward the Democratic party and toward Woodrow Wilson. As candidate for governor of New Jersey, Wilson had transformed himself from "a fierce partisan of the open shop" (as he earlier described himself) into a man who, in his own words, had "always been the warm friend of organized labor." When he became president, through acquiescence or affirmative aid, he helped unions secure a workers' compensation bill for federal employees, restrictions on child labor, legislation to improve the conditions under which merchant seamen worked, an eight-hour day for railroad workers, and a change in the anti-trust laws that seemed to exempt labor organizations. His secretary of labor, William B. Wilson, a former United Mine Workers of America official, set up machinery to mediate industrial disputes and helped the AFL's president gain access to administration officials.

The AFL leader, Samuel Gompers, sought cooperation with the federal government as part of a long-range plan to bring the state into labor's struggle for improved conditions and greater respectability. He had determined when fighting began in Europe that the AFL must support mobilization as a way of ensuring government support for union objectives. When the Council of National Defense was established, he took a seat on it alongside the industrialists and professionals and attempted to reconcile the interests of capital and labor. After the United States became a belligerent, he announced that the war was "the most wonderful crusade ever entered upon by men in the whole history of the world."

Concentrating on his long-term goal of labor participation in governing the economy, Gompers urged the AFL unions, made up of craftsmen whose skills were vital to the war, to avoid strikes in war industries. He declined to use the country's need as a weapon to abolish the open shop, a very generous concession since unions are considerably easier to ignore or eliminate in shops where all workers do not have to be members.

As part of Gompers's bargain with the administration and to gain more members for its own unions, the AFL battled against radical labor organizations, including antiwar socialists and the Industrial Workers of the World. As Gompers told the president, "Either the government and the employers generally will have to deal with representatives of the bona fide organized constructive labor movement of the country, or they will have the alternative of being forced to take the consequences of the so-called I.W.W." Through the CPI-financed American Alliance for Labor and Democracy, the AFL fought to gain worker support for the administration's foreign policy and battled the IWW and the People's Council. In return, the AFL received official recognition; government support for collective bargaining and for prevailing union wages, hours, and working conditions;

and federal assistance in its competition for members with the IWW. For instance, when the Department of Justice raided IWW offices, the government turned seized records over to Gompers.

Some labor leaders, including other members of the AFL, wanted to emphasize immediate benefits over Gompers's vision of long-term possibilities. "Now I want you to get it into your heads," James O'Connell of the Metal Trades told the Boilermakers' convention in September 1917, "to talk about dollars, not pennies. . . ." More aggressive union officials felt the president of the AFL was giving away the only weapons that organized labor had for making permanent inroads in American companies—the strike and the closed shop—in exchange for gains that a change of administration or the end of the war could sweep away. The head of the carpenters' union, William L. Hutcheson, refused to go along with any agreement that acquiesced in an open shop.

Still, conservative unions and social-justice reformers secured several advantages from the war welfare state, at least in the short run. Social workers helped promote their recently established profession and establish its utility by advocating programs to help America win the war. Labor unions sent representatives to some of the government boards that awarded contracts, and labor leaders joined members of the nation's business and professional elite on committees of the Council of National Defense and the War Industries Board. In return for what amounted to a no-strike pledge by the AFL and the agreement of Samuel Gompers not to seek permanent gains for AFL members, the National War Labor Board tried to shield workers from lockouts and to protect union organizers from employer interference—something no federal agency had ever done before. The Railroad Administration formally recognized the rights of railroad workers to bargain collectively and granted higher wages and improved working conditions. The War and Navy departments and the Emergency Fleet Corporation all prescribed detailed labor standards.

The War Department's labor codes suggest how numerous and detailed these standards could be. Every company making army uniforms had to pay minimum wages, maintain an eight-hour work day, permit inspections of sanitary facilities, comply with the federal Child Labor Act and all state labor legislation, and allow workers to bargain collectively. Every munitions manufacturer that contracted with the Ordnance Department was required to give its employees a five-and-a-half-day work week, pay them time and a half for overtime and cost-of-living increases, and provide decent sanitary facilities and comfortable heating and ventilation. They were forbidden to use sweatshops or hire children, and they had to provide women workers with special conditions: seats with backs, rest periods, a place to eat and sufficient time for meals, no lifting of weights greater than twenty-five pounds, no more than eight hours of work a day, and no work at all after sundown.

Contractors sometimes found ways of getting around these rules. To avoid genuine collective bargaining, some employers organized company unions and employee shop-representation committees actually controlled by management. Several employers complained about federal rules, and a few openly defied the government. But with government paying labor costs on cost-plus contracts and with the possibility, however remote, that it might seize companies that did not cooperate, most businesses fell into line. Unions grew by nearly 70 percent between 1917 and 1920, adding more than two million members, and in some industries where workers were already organized before the war, they were able to take control of shop committees and secure bargaining rights, at least informal union recognition, and sometimes even a union shop.

The administration did not grant or enforce labor benefits uniformly or give unions every concession their leaders asked for. Acting as a broker between management and labor, government responded to the strengths and weaknesses of both parties and to its dependency on each of them. It granted its largess to powerful and well-organized unions, like the railroad brotherhoods, more readily than to feeble ones. It paid serious attention to the demands of telephone workers only after they had strengthened their organization. It permitted company unions and allowed owners to have no unions at all where none had existed before the war. While the administration approved the eight-hour day in principle, it refused to make it mandatory. And it never required some of the important war agencies to treat union representatives equally with business leaders. One reason why so many contracts went to nonunion enterprise was the exclusion of union officials from contracting committees of the Council of National Defense. On the War Industries Board, as Grosvenor Clarkson observed, the AFL official sat not to represent labor but to manage it.

Yet even if workers' representatives were sometimes regarded as tokens and though some of the gains their unions made were limited and uneven, the idealism of reformers, the actions of laboring people, and the needs of a war government brought many workers gains and opportunities they had never previously enjoyed. Government programs to provide housing for war workers illustrate this phenomenon.

The war created a very serious housing shortage in the United States—consuming supplies needed for building, shifting workers out of the construction industry, and diverting to military production capital that might have been used to finance homes and apartments. Conditions near war plants and shipyards were especially bad. Rents rose spectacularly where it was possible to rent at all. Workers signed up for jobs, traveled to the plant

site, and a few days later, because there was no place to sleep, left. Since
the demand for labor was growing, workers did not have to accept these
conditions; yet the private-housing industry could not be expected to fill
their needs. What sane builder would put up homes near a place like the
Hog Island shipyard, located in the middle of nowhere, for people who
would leave the area as soon as their war jobs ended?

It should have been obvious before 1917 that miserable or nonexistent
housing would make it hard to turn out ships and guns. The British had
already learned this lesson and had started building housing for workers.
However, the United States had no public housing tradition and little
experience with planning towns for workers. The businessmen who were
trying to manage the war economy thought private enterprise could solve
the housing problem with a little help from the Treasury. Thus, a committee of the Council of National Defense recommended that the federal
government lend money at low interest rates to war contractors, who
would use it to build homes for their employees.

The military services and other war agencies decided that the government would have to intervene directly. The Ordnance Department began
to build housing for munitions workers. The Labor Department formed a
United States Housing Corporation to furnish living quarters to war workers while the Emergency Fleet Corporation started programs to house
shipyard employees and their families.

At first, the public corporations were reluctant to compete with private
builders. They tried to place workers in existing quarters and sought to
improve public transportation so that people could commute to distant
war jobs. But eventually they concluded that they would have to construct
new housing at government expense. The Emergency Fleet Corporation
hired real-estate companies that were subsidiaries of the shipyards to build
housing projects on land that the yards supplied. It chose the architects
and the landscapers and kept control over rentals, sales, and management.
The EFC provided over 8,000 houses and nearly 850 apartments for
families and accommodated 71,500 single employees in dormitories,
boarding houses, and hotels. The United States Housing Corporation,
meanwhile, bought its own land, planned its projects, built them, and
managed them, constructing or locating living quarters for nearly 6,000
families and over 7,000 single persons. Because the government programs
began so late in the war, the houses they built could not be occupied until
after the Armistice.

Some wartime government projects strongly emphasized aesthetic
quality, a departure for working-class housing and an anomaly in the
history of public housing in United States. Before the war, most American
housing reformers concerned themselves with matters that could be quan-

tified, such as room size, number of windows, the amount of ventilation, and the number and location of toilets. The wartime agencies tried to improve intangible as well as measurable living conditions. Their architects, designers, and town planners included highly talented and even visionary people who conceived their projects as experiments for the housing of the future. E. D. Litchfield, chief architect of the Emergency Fleet Corporation's Yorkship Village, described his project for war workers in Camden, New Jersey, as a

> romantic opportunity . . . for the Government to produce an industrial community which should be, as far as reasonable economy and the urgency of the case would permit, an example to private enterprise throughout the land; which would show how, through providing proper homes for its employees, an industrial corporation could lay the foundation for a contented and efficient body of workers.

Some of the wartime projects were drab and ill-proportioned, laid out in monotonous gridirons. Several were models of simple beauty whose designers arranged streets to follow the natural contours of the land and integrated parks, lawns, trees, and squares into their developments. In the better projects, houses were orderly and efficient, yet harmonious, constructed on standardized plans but varied through landscaping, siting, and the use of different kinds of roofs. Where the area itself was attractive, planners tried to fit buildings to their surroundings, making a project on Long Island, for example, resemble a traditional Long Island fishing village.

Because they understood that workers and their families who came to a remote area would need places to shop and socialize, government planners designed some of the projects as communities. The Ordnance Department's housing branch oversaw the building of one of these communal developments in Perryville, Maryland, for employees of the Atlas Powder Company. Perched on the high banks of the Susquehanna River, it consisted of three hundred white, wood, detached homes with green blinds and grey-green roofs, together with a club, a store, a theater, and a school. Union Park Gardens, an Emergency Fleet Corporation project in Wilmington, Delaware, included sites for a school, a playground, and a community building containing pool rooms, a gymnasium, a room for smoking, sewing and reading rooms for women, and a six-hundred seat auditorium.

Before the war and in later years, private builders developed housing communities like these for affluent citizens of American society. During the Great War, as part of its program for mobilizing the nation, the American government built beautiful harmonious housing for ordinary working people.

At the same time and also to ensure that the war for democracy did not leave a large part of its people destitute, the federal government provided millions of servicemen and their families with social-insurance benefits. Members of the armed forces became eligible for government medical care and were allowed to purchase cheap government life insurance. As a result of legislation that social-welfare reformers helped to write, the government paid allowances to dependents of enlisted men and gave death benefits to the widows and orphans of those who died for their country. To disabled servicemen, it provided monthly payments, guidance and job retraining, and a variety of other benefits. All this cost the United States a good deal of money. By 1962 the Veterans Administration and the agencies that preceded it paid out more than thirty billion dollars for expenses stemming from World War I.

State governments, meanwhile, were expanding their own social-welfare programs. Several enacted new minimum-wage, maximum-hour, and workmen's compensation laws and tightened existing legislation, adding new safeguards for women and children. During the year ending September 30, 1917, eleven states passed stricter child-labor laws; four inaugurated child-labor legislation; six approved compulsory education bills; and five instituted pensions for mothers. Nevada and New York limited the hours that women could work in certain occupations. Washington helped working mothers by establishing free kindergartens. Tennessee abolished the contract labor system. Arizona provided for the location of child-welfare boards in each of its cities, and Wisconsin's industrial commission prohibited night work in laundries and manufacturing establishments. Though businessmen presented the Massachusetts industrial commission during one period in 1917 with thirty-two requests to permit night and overtime work for women, the commission granted only a fourth of these, and it allowed only one of thirteen requests to have minors work nights or overtime.

Some of these actions and decisions came too early to reflect the full impact of the Great Crusade and may therefore be considered a residue of a prewar reform movement. Yet in 1918, several states actually strengthened their welfare or labor programs. Virginia enacted a workmen's compensation law, while Kentucky and Massachusetts liberalized existing legislation. New York forbade women under twenty-one to work in certain dangerous trades and, along with Virginia and New Jersey, strengthened child-labor prohibitions. Rhode Island outlawed a hazardous textile machine. New Jersey established an employment and migrant welfare bureau while strictly regulating private employment agencies that exploited migrant workers from the South.

Wartime social-justice programs did not bring equal advantages to all groups of Americans. The special protections for women in the workplace encouraged employers, after the troops returned, to replace women with men, who could legally work long hours under dangerous conditions. Some people actually lost ground. Families with a number of wage or salary earners had better opportunities to raise their living standards than those supported by a single breadwinner. While highly skilled workers in war industries substantially increased their real income, wage workers as a group made only slight gains, and government employees lost more than 23 percent of their buying power between 1915 and 1918.

Many of the wartime advances did not survive after the fighting ended. The housing projects were sold to private companies. Employers openly defied the later decisions of the War Labor Board and rolled back earlier ones. After the Armistice, with Congress under Republican control and the Wilson administration unable and unwilling to coerce employers and afraid of radicalism and industrial violence, labor ceased to benefit from federal intervention. When the administration did intervene in a postwar steel strike, it was by sending troops who helped stop a union-organizing drive. American corporations quickly proved how one-sided Gompers's bargain had been by ridding factories of unions and labor organizers.

The arrangements that social reformers made with the Wilson administration also were short-lived. During the war, reformers had imagined that if they cooperated with an administration that suppressed dissenters and crushed revolutionaries, or if they at least kept still, government would employ what they called "social control" to build a better society. But as radical critics* observed at the time, social control could mean many things, including domination and manipulation by a government utterly opposed to social reform.

For a while, though, reformers and labor leaders who ignored the dangerous possibilities and had not yet grasped how temporary the reforms would be, thought wartime advances worth the cost. In May 1919 a writer observed in *Survey*, a social workers' journal, how

> hundreds of thousands of soldiers' families are being safeguarded if the
> need arises, in accordance with the principles of family rehabilitation
> worked out by charity organization societies everywhere. Conditions of work
> in munitions plants and under government contracts are responding to the

*Such as Randolph Bourne (see Chapter 8).

standards long advocated by consumers' leagues and other similar
bodies. . . . Industrial housing has turned in part to the established housing
movement for instruction. Social insurance, the care and education of the
blind and crippled, employment methods—these and a dozen other aspects
of national well being are undergoing similar transformations.

In the "most terrible and disastrous of all wars," social-justice reformers
and organized workers friendly to the government had achieved an
ephemeral triumph.

6

The Great War and the Equality Issue: African-Americans and Women

The Wartime Campaign for Racial Equality

When America began its crusade to free the victims of German autocracy, millions of its own black citizens, oppressed by poverty, segregation, political impotence, and violence, were badly in need of liberation. Yet even before the United States became a belligerent, the Great War had begun to change their lives. War accelerated a migration of rural blacks to the cities, particularly to large industrial cities in the North. When factory owners found themselves unable to import European workers to fill Allied war orders, they sent recruiters to the South, where many black inhabitants, plagued by floods, crop failures, and discrimination, were ready to tear away from their roots. Pressed by conditions at home and drawn by reports of good wages and better living conditions, tens of thousands moved northward. Between 1910 and 1920, the black population of Chicago grew from about 44,000 to over 109,000. Government statisticians estimated that over 153,000 African-Americans lived in New York City in 1920 as compared with less than 92,000 a decade earlier. Altogether, the black population of the North and West grew by some 333,000 in those years.

As African-Americans left rural areas, violence pursued them. The number of lynchings grew during the war years. In several cities of the North and South, racial clashes broke out in 1917, including a series of riots in the spring and summer when thousands of people surged through the streets of East St. Louis, Illinois, burning, looting, and killing. In late August, provoked by racial incidents, including the beating of a black woman by a white policeman, hundreds of black soldiers of the 24th Infantry Regiment stormed into the nearby city of Houston. They assaulted the police station, killing fifteen whites or hispanics, including five policemen, while losing four of their own men. In the courts-martial that followed, sixty-four soldiers were tried, twenty-nine received the death penalty, five were acquitted, and the rest received severe prison sentences. Of those condemned to die, nineteen were hanged, and the president commuted the death sentences of ten. Shocks emanating from these events affected both the way African-Americans reacted to the war and efforts of the United States government to manage blacks in wartime.

In the prewar years, black leaders had developed differing strategies for dealing with racial oppression. Followers of Booker T. Washington, the founder and head of Tuskeegee Institute, believed the wisest policy was to accept the color line, at least publicly, and to seek opportunities for economic advancement within a segregated system while quietly working to reduce the harshness of America's racial order. Others like the sociologist and historian William E. B. Du Bois demanded rapid change, insisting that blacks should accept nothing short of full equality in all areas of life.

The accession of Woodrow Wilson to the White House offered little hope to the advocates of racial equality. While Wilson considered particular African-Americans worthy individuals, this southern-born president had little regard for black people in general. In the 1880s, he had referred to them privately as an "inferior race" and doubted they should be allowed to vote, although he later came to believe that in two or three centuries blacks would probably secure some racial and political equality. While he advocated compulsory education for black children, as president of Princeton he continued a policy of excluding black students. Wilson found it hard to contemplate social equality between blacks and whites, which he feared would lead to intermarriage and degrade the white nations. In 1915, when a niece of Edith Bolling Galt became engaged to a Panamanian, Wilson remarked, "It would be bad enough at best to have anyone we love marry into any Central American family, because there is the presumption that the blood is not unmixed; . . ." A Princeton alumnus recalled Wilson as "the best narrator of darky stories that I have ever heard in my life." In a love letter to Mrs. Galt that year, he described himself as "free, white, and twenty-one."

Wilson chose for his cabinet Secretary of State Robert S. Lansing, who believed that people of African ancestry were incapable of self-government, and Postmaster General Albert S. Burleson and Secretary of the Treasury William G. McAdoo, both of whom segregated blacks in their departments. Besides greatly extending Jim Crow rules in the federal government, Wilson's administration gave whites patronage appointments that his predecessors had reserved for blacks. When people protested, Wilson informed them that segregation was in the best interests of both races.

This kind of answer did not satisfy the African-American editor

William Monroe Trotter. When Trotter complained bitterly to the president about the unequal treatment of blacks, Wilson replied in an angry
voice, "You are an American citizen, as fully an American citizen as I am,
but you are the only American citizen that has ever come into this office
who has talked to me with a background of passion that was evident."
Then he showed Trotter the door.

Despite the administration's record, some African-Americans believed
they could use the war to persuade the federal government to help them.
The National Association for the Advancement of Colored People, an
organization of blacks and white liberals, tried unsuccessfully to have
Congress outlaw lynching—as a war measure. Many blacks hoped the
country would lower its racial barriers as a reward for their wartime loyalty
and sacrifices. A few offered an explicit quid pro quo—for blacks to
support the war enthusiastically would depend on the administration's
racial policies. Thus, when the Los Angeles Forum, an African-American
community action organization, sent the president a message asking him to
end discrimination in the army and navy, to protect blacks against out-
breaks of violence (like the East St. Louis riots), and for help in desegre-
gating railroads, cafes, and hotels, it declared, "if the country will not
protect us in time of need, we feel that it is not humanly possible to
nourish our hearts to loyalty by the memory of cold neglect from [the]
general government."

The most militant and radical blacks, younger writers and orators,
socialists, and black nationalists, were less hopeful and less interested in
coming to terms with Wilson's government. They recalled that African-
Americans had served in earlier conflicts only to face continued rejection
and discrimination. They noted how even in the midst of the government's
crusade to liberate foreigners whites continued to terrorize American
black people and kill them in race riots. Some radicals concluded that
African-Americans should take no part at all in the war, that it was a white
man's conflict.

This was a dangerous stand for anyone to take in wartime America.
"The bold truth," poet James Weldon Johnson insisted, was that "the
Negro cannot afford to be treated as a disloyal element in the nation.
Imagine the results if he should arouse against him the sentiment which is
now directed against the pro-German element." Nevertheless, militants
expressed their desire for change and their distrust of the administration in
newspapers (such as the *Messenger,* edited by A. Philip Randolph and
Chandler Owen), and in demonstrations. Thousands of African-Americans
marched down Fifth Avenue in New York City on July 28, 1917, to protest
racial violence and discrimination. Their signs read, "Make America Safe
for Democracy," "Taxation Without Representation Is Tyranny," and

"Mother, Do Lynchers Go to Heaven?" One poster, which the police persuaded the marchers not to display, showed a black woman kneeling before President Wilson and appealing to him to make the United States safe for democracy before trying to do so for Europe.

The Wilson administration needed the labor of black workers, particularly in agriculture, and of the hundreds of thousands of black men it was drafting into the army. These signs of disaffection among African-Americans, particularly the indications of poor morale and rebelliousness among African-American troops, troubled the president and his advisors. Unwilling to address the basic causes of the problems, the administration attempted to manage African-Americans much as it was managing white workers. It appointed conservative blacks to positions with more visibility than power and tried to use them to communicate with and control other African-Americans. It disseminated propaganda to black audiences; and it suppressed black radicals who insisted on saying how they felt about the war.

In August 1917, the month of the Houston uprising, the War Department named Emmett Scott, who had been Booker T. Washington's secretary, as special assistant to the secretary of war on issues affecting African-Americans. Scott attempted to improve conditions for black servicemen. Meanwhile, the administration employed him to bolster their morale and to build enthusiasm among black civilians. Scott helped the CPI establish a Committee of 100 Citizens, made up of prominent black journalists, lawyers, clergymen, educators, and political appointees, the task of which was to stimulate the patriotism of African-Americans.

The CPI arranged for black ministers and community leaders to deliver speeches promoting the nation's war aims and encouraging loyalty. Its appeals and the messages of other government agencies offered black people reassurance, promises, and warnings. Through Emmett Scott, the CPI told them that America was fighting a "people's war . . . not a white man's war . . . not a black man's war," but "a war of all the people under the Stars and Stripes for the preservation of human liberty throughout the world." It was a war in which American blacks had a very large stake, Scott explained, for if the Germans won and enslaved the American white man, the ultimate consequence would be the "reenslavement" of the American black man and the loss of "all that he has gained since the Emancipation Proclamation." One CPI bulletin promised that the war would lead to "a wonderful amalgamation of the races within America."

While leaders like Scott were announcing publicly that black America was united behind the war, they were very much aware of contrary indications and told white officials that serious problems had arisen among African-Americans. Nearly a year after his appointment, Scott was so disturbed about unrest among blacks that he urged the women's divisions of the Council of National Defense and of the Committee on Public Information to take steps against dissent among black women in the South. In June 1918 George Creel, alarmed about "rumor and ugly whisperings," including stories prevalent among African-Americans that black troops were being sent into the most dangerous areas to spare the lives of white soldiers, had Scott call a secret meeting of high government officials with black newspaper editors and other black leaders. At this conference, which began in heated controversy, Secretary Baker denied and attempted to disprove some of these stories. Baker and Creel wanted the black press to write more positively about the war, but the immediate response of the editors was a message to the president warning him that the justifiable grievances of black Americans had created bitterness and unrest and that "the apparent indifference of our own Government may be dangerous." Nevertheless, they agreed to cooperate.

While it tried to influence African-American opinion through the news media, the administration was taking steps to silence black dissidents. Treating them like other groups whose loyalty it questioned, it placed outspoken protestors under surveillance, pressured them to alter their views, and jailed those who refused to keep still. Blacks working for the Department of Justice were sent to listen in at meetings of the National Association for the Advancement of Colored People. Black informants in African-American army units kept track of dissenters for the Military Intelligence Division of the War Department. Other federal agents monitored the activities of blacks associated with the Socialist party and infiltrated black organizations suspected of opposing the war.

African-American publications that the government regarded as inflammatory were suppressed. The Justice Department repeatedly warned W. E. B. Du Bois about articles he was publishing in the *Crisis*, the NAACP journal, and the Post Office held up an issue that contained documents illustrating discrimination in the army. Agents of the Bureau of Investigation arrested A. Philip Randolph and Chandler Owen during a socialist-sponsored antiwar tour, charged them with violating the Espionage Act, and inquired into their draft status. The Post Office suspended mailing privileges of the *Messenger* and other African-American papers for criticizing the administration or for publishing critical letters. The editor of the *San Antonio Inquirer* was given a two-year sentence for printing a letter by Mrs. C. L. Threadgill Dennis, who told the Houston mutineers that while black women regretted they had mutinied and shed innocent blood, it was better that the mutineers be shot for trying to protect a black women than to die fighting to make the world safe for a democracy blacks could not enjoy.

Threats and repression had some effect. "Everybody watch his tongue," warned the *Baltimore Afro-American*, which is what organizations like the NAACP did. At a June 1918 meeting, its editorial board told Du Bois to present only "facts and constructive criticism" in the *Crisis*. In what may have been a coincidental burst of optimism, Du Bois produced a series of pro-war editorials including one that declared:

> Let us, while this war lasts, forget our special grievances and close our ranks shoulder to shoulder with our own white fellow citizens and the allied nations that are fighting for democracy. We make no ordinary sacrifice, but we make it gladly and willingly with our eyes lifted to the hills.

In response to persuasion and encouragement, as the result of spontaneous patriotism, or in the hope of turning the war into a campaign for racial democracy at home, millions of African-Americans supported the United States government in World War I, as their ancestors had done in earlier struggles. Black clergymen led bond drives and urged eligible men in their congregations to join the armed forces. Black colleges threw themselves into the war effort. Howard University did away with German-language courses. Blacks purchased hundreds of millions of dollars' worth of Liberty Bonds and stamps and took hundreds of thousands of war jobs and staged loyalty parades under the auspices of the Council of National Defense. The vast majority of African-American newspapers spoke favorably of America's part in the war. The NAACP encouraged blacks to participate in Liberty Bond drives, helped organize food-conservation programs, and cooperated with draft boards. Even William Monroe Trotter's organization, the National Equal Rights League, grudgingly backed the war effort. Some four hundred thousand black men and women served in the nation's armed forces.

While many of these servicemen joined with the same patriotic emotions that animated large numbers of white troops, some appear to have hoped that the nation would reward blacks with a measure of racial justice if they offered up their lives. That expectation proved illusory. The navy confined its slightly more than fifty-three hundred African-Americans to low-level noncombat jobs, mostly as stewards. The U.S. Marine Corps excluded blacks altogether. The army, which began the war with over ten thousand regular black troops, maintained a rigid color line. Recruiting officers turned down black volunteers, allowing only four thousand to enlist while six hundred fifty thousand whites were joining up. At the same time, the country's all-white draft boards conscripted blacks in disproportionate numbers. Instead of commissioning black doctors and dentists, the army drafted them as privates. It excluded blacks from the pilot section of the AEF Air Service. Most blacks were never allowed to prove their loyalty on a battlefield but were limited to menial work behind the lines. Nearly 90 percent of black troops, regardless of their qualifications, were required to work as laborers in the Services of Supply.

While the War Department and the American officer corps contained many individuals who despised black people, the secretary of war was probably the most "tolerant" member of Wilson's cabinet on racial issues.° Baker thought of black Americans as part of an underdeveloped race who, by acquiring education in the industrial arts, were "rendering themselves more useful in our civilization and more entitled to our respect." He told Emmett Scott that his official policy was to discourage racial discrimination against any person because it was not just and because he wanted to avoid disorder. But he also declared that for the duration of the war he had no intention of settling the "so-called race question."

When Emmett Scott relayed protests against compelling black soldiers to serve as laborers rather than combat troops, Baker dismissed these grievances with the remark that there was "far less hazard to the life of the soldier connected with the Service Battalion than on the firing line." (He was thinking perhaps of the rumors that black troops were being sacrificed to spare the lives of whites.) To the complaint that blacks could not enter all-white arms like the AEF Air Service, the secretary replied that he intended to preserve the "custom of the army," which had been to place "colored people into separate organizations." Following this tradition, his department confined African-American combat troops to a Ninety-second (colored) Division and four additional black regiments, three of them made up from National Guard Units.

The commanders of all these units and most of their officers were white, following the supposition that black troops would fight well only under the leadership of Caucasians. But the army had other reasons for not wanting African-Americans in the officer corps. General Leonard Wood had excluded them from citizens military training camps when he was chief of staff, saying that he did not wish to admit the kind of people "with whom our descendants cannot intermarry without producing a breed of mongrels." The commander of the 372nd Infantry Regiment requested that Caucasians replace black officers in his unit, partly on the ground that racial distinctions were "a formidable barrier to . . . that feeling of comradeship" essential to mutual confidence and esprit de corps. During the First World War, the idea of sharing quarters and mess with

°The War Department official in charge of racial matters, Under Secretary Frederick P. Keppel, later sponsored Gunnar Myrdal's pioneering study of America's race problem, *An American Dilemma*. Baker had suggested this project.

black officers, of saluting them and deferring to them if they bore higher rank, and of fighting by their sides was unthinkable for many white American officers.

When the United States entered the war, its highest-ranking black officer was Colonel Charles Young of the Tenth Cavalry, the third African-American to graduate from the United States Military Academy and the only one still in uniform. Young had served as a military instructor at an Ohio university, performed noncombat duty in the war with Spain, and in 1916 led black cavalry troops against Pancho Villa's guerrillas. Despite his record, several white officers did not wish to serve under him, and four of them made that point in letters to their senators. The senators relayed their complaints to the secretary of war and the president, who told Baker he feared "serious" trouble if Young remained in his command. The colonel was sent to an army hospital for a physical examination, and when the examination disclosed that he had high blood pressure, the War Department retired him as disabled. Later, Young rode to Washington on horseback from his home in Ohio to prove that he was fit, but to no avail. Black newspapers made an issue of Colonel Young's case. Robert R. Moton complained about it to the president and asked him to keep Young in the army. Wilson replied that there was "no possible ground" for the belief that Colonel Young was being discriminated against in any way.

Black college men who asked for an opportunity to earn commissions met frustrating resistance. The army would not allow them to train in integrated units, yet Congress had made no provision for segregated officer candidate schools. Students and teachers at Negro colleges and universities protested. Black leaders suggested to the president that if his administration did not set up at least a segregated officers training camp, ten million African-Americans might not feel the "active and healthy patriotism" the government desired. Finally, after pleas from the NAACP and the president of the Committee of 100 Citizens, the War Department agreed to set up a special camp for black officer candidates at Fort Des Moines, Iowa.

Some eleven hundred African-Americans received commissions, 1 percent of the officer corps of an army with 13 percent black enlisted men. But even after they had pinned on their bars, many still did not receive the normal formalities of military courtesy. Charles H. Williams, an investigator for the Federal Council of Churches reported that when black officers first appeared at Camp Funston, nearly 90 percent of the whites refused to salute them. Sergeant Sydney A. Gaines, a white noncommissioned officer explained why. He wrote in his memoirs about the day he was entering new quarters at Camp Pike, Arkansas, when "low and behold on the same day we moved in, a bunch of nigger cadets moved right in next to us, commanded by nigger officers. All of us were from Texas and we were not about to get used to saluting nigger officers."

Many white officers shunned their black counterparts, and some of them tried to discredit black colleagues in front of soldiers the blacks were expected to lead. When the officers of a white cavalry troop were asked to sign a document stating if they would be willing to command black enlisted men, they talked it over and agreed to do so; but one of them, Lieutenant Milton Bernet, wrote his mother, "of course [we] would not serve with negro officers. Heaven forbid!"

Army policy prevented all but a tiny number of blacks, however qualified, from ever rising above captain. The adjutant general issued instructions excluding African-Americans from assignments that could bring them higher rank. When Emmett Scott inquired at the War Department why a particular black officer in the Ninety-second Division had not received a promotion despite the recommendation of the division commander, the adjutant general referred Scott to these instructions, adding that he found no evidence of discrimination.

The question of where to train black units was highly sensitive and became even more so after the Houston uprising. Southern whites did not want armed blacks in their midst, particularly men from above the Mason-Dixon line. When J. F. Floyd, the mayor of Spartanburg, South Carolina, heard that the Fifteenth Infantry, a black National Guard unit from New York City was coming to his town to train, he declared that he was sorry to hear it for he expected that the Guard troops "with their northern ideas about race equality . . . will probably expect to be treated like white men." But Floyd made it plain that "they will not be treated as anything except negroes." Senator Vardaman of Mississippi warned that conscripting African-Americans would put "arrogant strutting representatives of the black soldier in every community." In several communities in the South, local people made no attempt to hide their dislike of African-American soldiers and tried to make them conform to Jim Crow customs.

Civilians who worked with the army as auxiliaries helped enforce the color line. The Federal Council of Churches investigator Charles Williams reported that some YMCA secretaries were even more prejudiced than army officers. Williams estimated that most "Y" secretaries served blacks half-heartedly or made them feel unwanted when they served them at all. At Camp Greene in North Carolina, Williams found five YMCA buildings, none of them for the ten thousand black troops stationed there.

Racial discrimination affected African-American troops outside the South. When a black sergeant stationed with the Ninety-second Division decided to integrate a Jim Crow theater in Manhattan, Kansas, the commanding general, Charles C. Ballou, reprimanded him. Ballou, a New

Yorker who had led black troops in wars against the Indians and the Filipinos, admitted that the sergeant was within his legal rights. He even pressed charges against the theater operator. But he issued a bulletin reminding the men of the Ninety-second that he had urged them not to go where their presence would be resented, and he warned them against any action, "no matter how *legally* correct," that would provoke race animosity. The success of the division, he declared, "with all that that success implies" depended on the goodwill of the public. The public was nine-tenths white. White men had made the Ninety-second Division, Ballou declared, and could break it just as easily if it became a troublemaker.

Black servicemen reacted in varying ways to racial hostility and to the abysmal conditions they encountered in training camps like Camp Hill in Newport News, Virginia, where some of them spent the frigid winter of 1917–1918 in floorless tents without stoves or blankets and with no changes of clothing or bathing facilities. Some ignored these problems or enjoyed their stay in the army in spite of them, hid bad feelings, or became or pretended to be apathetic. They griped to one another, and they expressed how they felt in songs:

> Jined de army fur to git free clothes,
> What we're fighting 'bout nobody knows:

Thousands complained through military channels about local instances of discrimination and about the wretched conditions they encountered. Occasionally, they did more than complain. After a series of racial incidents, the New York National Guard unit in Spartanburg appeared ready to riot. Emmett Scott rushed down from Washington to plead with them not to act rashly. The War Department hurriedly shipped them to France.

When the American Expeditionary Force left for Europe, the caste system went along. This was partly the result of long-ingrained attitudes. Milton B. Swenigsen, a private from the Midwest wrote his mother:

> Saw, yesterday, a real American negro who is with a work battalion over here. It may seem odd but nevertheless it seemed good to hear even a negro talk again. And he would pass for one of the "black face" comedians of the stage when he was telling his experiences over here.

Lieutenant Bernet wrote home about the way American black troops had not "forgotten their little mannerisms" and how hundreds of them, working in road gangs, were "singing coon songs and revival songs in their peculiar shrill crescendo." When an officer in France complained to the

Stars and Stripes, the army newspaper, that the only things it printed about American black soldiers were "ridicule, degrading remarks and prejudicial propaganda," the editor replied that the complainant was "altogether too sensitive," his point of view was "warped," and far from hurting blacks, "the stories of colored soldiers, told in Negro dialect . . . make us have a larger liking for the colored man, . . ."

Calculation as well as traditional racist attitudes conditioned the treatment of African-American soldiers in France. The French were notoriously color blind, and American whites feared that if Frenchmen (or worse, French women) treated black soldiers as ordinary people, it would be impossible to keep them in check when they returned to the United States.

The AEF had no official race-relations policy, but its members took unofficial steps to prevent the French from associating with blacks or treating them as equals. White doughboys informed French women that black men had tails and horns. A French liaison to AEF General Headquarters, Colonel J. L. A. Linard, conveyed to French military and civil leaders how the American command wanted them to treat blacks. Linard explained that white Americans considered colored people a lower class of human beings and were deeply concerned about French indulgence and familiarity with them. He urged French officers not to praise American black troops too highly, especially with American whites present, to allow expressions of intimacy between black soldiers and white women, and he asked local officials to keep the French people near the army bases from spoiling the blacks.

White officers instituted Jim Crow rules. Signs appeared on buildings saying "Whites Only." At Saint-Nazaire, black stevedores were told to stay out of French homes and not to be seen with French civilians. For weeks, the commander of a segregated supply depot confined black enlisted men to the depot and base area while white troops went to a nearby town on pass. Several blacks defied the rule by going absent without leave, got into trouble with the military police, and ended up in the guardhouse. Finally, another officer took charge of the demoralized troops, restored their right to visit nearby towns, personally pulled down some of the Jim Crow signs, and banned the use of the word "nigger." Within three weeks, the number of men in the guardhouse dropped from three hundred to fifty.

African-American officers were subjected to special forms of racial harassment. When the troopship *George Washington* left for Europe, white officers of the 368th (colored) Regiment received first-class staterooms while its black officers were assigned to second class. The lone black officer on a ship returning to the United States had to eat by himself after about four hundred whites had been fed. Charles Houston, dean of the Howard University Law School, who had served as a lieutenant in the Ninety-second Division recalled:

> The hate and scorn heaped upon us as Negro officers by Americans at Camp Mencou and Vannes in France, convinced me there was no sense of dying in a world ruled by them. . . . They made us eat on benches in order to maintain segregation and they destroyed our prestige in front of French officers.

Some blacks and sympathetic whites believed the army wanted African-Americans to fail in battle. This suspicion derived from the way black officers were treated, from the fact that the four regular-army black regiments were not sent to fight in Europe, and from the poor equipment and training given to black soldiers who did go to France. The Ninety-second Division entered combat gravely unprepared. It lacked essential equipment. It had trained in seven different places, sometimes under the supervision of inexperienced junior officers and noncoms. A quarter of its enlisted men, who joined it during the month before it left for France, had almost no training at all. In France, members of the division received pre-combat instruction for no more than a week and in some cases for just a few hours.

Some of the hastily mobilized white divisions had similar weaknesses in training and equipment, but the Ninety-second also faced racial hostility and discrimination that demoralized enlisted men and hampered its black officers. Capable officers who wanted to stay in the division were transferred out, and incompetent officers, white and black, were transferred in, so that, in the words of its commanding general, the Ninety-second became a "dumping ground" for undesirables. Between white and black officers, racial friction never let up. "It was my misfortune," General Ballou later wrote, "to be handicapped by many white officers who were rabidly hostile to the idea of a colored officer, and who continually conveyed misinformation to the staff of the superior units, and generally created much trouble and discontent." Some black junior officers believed that whites had given commissions to black officers not on the basis of merit but on the likelihood that the recipients would fail in combat. White officers believed their black counterparts had received commissions only as a result of political pressure. Enlisted men in the Ninety-second trusted few of their officers, black or white.

Most of the division's white regular-army officers, selected because of their presumed knowledge and experience with blacks (that is, because they were from the South), expected their troops to fight poorly. General

W. H. Hay, a brigade commander of the Ninety-second thought "the inherent weaknesses in negro character, especially general lack of intelligence and initiative," made it necessary to give blacks more training than whites to prepare them for combat. General W. P. Jackson, who commanded the division's 368th Regiment, had concluded before the war that blacks only fought well when led by officers who realized "that the colored soldier is really a grown-up child" and treated him accordingly. The commander of the 367th Regiment declared that "as fighting troops, the negro [*sic*] must be rated as second class material, this due primarily to his inferior intelligence and lack of mental and moral qualities." When Robert L. Bullard, commander of the Second Army, visited the division after the Meuse-Argonne offensive, he wrote that except for General Ballou, not one of its ranking officers believed the troops of the Ninety-second would ever amount to anything as soldiers. "Every one of them would have given anything to be transferred to any other duty. It was the most pitiful case of discouragement that I have ever seen among soldiers."

Ballou himself was convinced that environment and education had produced serious defects along with virtues in his soldiers, descendants of people in whom "slavish obedience and slavish superstitions and ignorance were ingrained. . . ." He remarked after the war that "the average negro is a rank coward in the dark." General Bullard's racial feelings were even stronger. "Poor Negroes!" he noted in his diary. "They are hopelessly inferior. . . . With everyone feeling and saying they are worthless as soldiers, they are going on quite unconcernedly." Bullard believed that black officers convicted of cowardly behavior in battle should receive clemency because as blacks, they could not be held as responsible as white men.

The collapse of part of the 368th Infantry Regiment, the only unit of the Ninety-second (except for two machine gun companies) that saw combat in the Argonne offensive, provided evidence for such conclusions. At the beginning of the battle, the First Battalion of the 368th gained considerable ground, even without hand grenades, wire-cutters, signal flares, and artillery support; but the Second and Third battalions were turned back twice, disintegrated, and ran. After this incident, General Ballou requested that thirty black officers, all recently commissioned, be court-martialed for cowardice. The Ninety-second Division's other regiments were held in reserve, and after the 368th's two battalions broke down, the whole division was taken out of combat.

Ballou perceived a sinister reason for withdrawing his troops. After the war, he charged that white AEF officers wanted to get the black troops out of the Argonne "while their credit was *bad* as many preferred it should remain." He recalled that a colonel of one of the Ninety-second's regiments had come to him during the battle and on behalf of its officers begged him to send the regiment to the front. "Those men would have been dangerous at that time," Ballou wrote, "and ought not to have been humiliated by being sent to the rear."

On other occasions, when it had a chance to fight, the Ninety-second fought well, and so, on the whole, did the black regiments that served with the French. One of the best regarded of these was the 369th Infantry, the new designation for the Fifteenth (New York) National Guard Regiment. Colonel William Hayward, an attorney from New York City, commanded the 369th, and its other officers, most of them white, were New Yorkers who had recently graduated from Ivy League universities or who had served in National Guard units in other states. Between them and the "Men of Bronze," as soldiers of this unit called themselves, there was considerable respect. The French thought highly of the 369th, which brought to France not only able soldiers but also a band that included some of America's leading musicians, including its bandmaster James Reese Europe and the renowned singer and composer Noble Sissle.

After several weeks of duty as labor troops, the 369th went to the front, where it spent six months with only two weeks' rest. During those months, it received hundreds of completely untrained replacements, many of them illiterate and with no attachment to the regiment, who diluted the regiment's morale and tended to vanish before and during battle. The fighting spirit of the 369th was also weakened, Colonel Hayward reported, by unrelieved drudgery in the trenches, lack of opportunities for training, and the refusal of Pershing's headquarters to respond to requests to cite members of the regiment for bravery. Nevertheless, the New York regiment distinguished itself several times on the battlefield, and the French awarded the Legion of Honor or the Croix de Guerre to a hundred seventy-one of its officers and men.

In their postwar appraisals of combat performance, white officers, with few exceptions (such as Colonel Hayward), found it difficult to give African-Americans their due. General Jackson, who had led the black regiment that broke apart in the Argonne, observed

> the racial feeling between the white and the colored man in the United States is so strong that it is difficult to obtain a true appreciation of the worth of the colored man as a soldier. If he fails, his failure is subjected to more severe criticism, his faults are magnified to a greater extent, and his acts are more quickly criticized than in the case of a white soldier.

Several regular-army officers, including the Ninety-second Division's regimental commanders, replied to questions from the Army War College about the performance of African-American troops in the war by citing the failure of the two battalions of the 368th Regiment in the Argonne offensive. They regarded that as proof that black officers could not be trusted to lead their men properly and that black enlisted men would not fight well unless led by whites. Few of the officers who corresponded with the War College analyzed the special difficulties that black units confronted. The critics did not mention that during this and other actions, tens of thousands of white-American troops straggled in the rear or became disorganized and ran away and that three of the nine white-American divisions that began the Argonne battle also suffered from panic and disorder, could not be regrouped, and had to be replaced. These divisions, like the Ninety-second, were led by inexperienced officers and had little training. Unlike the Ninety-second, they were later given opportunities to redeem themselves.

The record of African-American soldiers in World War I was not just distorted; it was lost from national memory, at least from the memory of most Americans. In the decades that followed, the ordinary white inhabitant of the United states learned nothing about the activities of the black regiments that served alongside the French or about the Ninety-second or the Services of Supply. To white America, the "Men of Bronze" were invisible. That was one of the reasons blacks never realized their hope of redemption through war service.

The Wilson administration achieved essentially what it wanted from African-Americans—not brilliant combat performance but the kind of labor at home and behind the battle lines that was essential to fighting the war. To African-Americans, the war brought hope but little fulfillment, and its aftermath brought deepening poverty for those who remained on southern farms, expulsion from wartime jobs, denial of promotion in the army, a reincarnated Ku Klux Klan, continued discrimination, lynchings and burnings, and in 1919 a whole summer of race riots.

Some black leaders who had chosen to cooperate with the Wilson administration and had promoted the war as a means for securing better lives for their people were bitter at the outcome. W. E. B. Du Bois declared in 1919:

> We stand again to look America squarely in the face and call a spade a spade. We sing: This country of ours, despite all its better souls have done and dreamed, is yet a shameful land.
>
> It *lynches*. . . . It *disfranchises* its own citizens. . . . It encourages *ignorance*. . . . It steals from us. . . . It insults us. . . .
> We *return*. We *return from fighting*. We *return fighting*.
>
> Make way for Democracy! We saved it in France, and by the Great Jehovah, we will save it in the United States of America or know the reason why.

The war's results were not completely negative for America's black citizens. The migration to the North, stimulated by the war, brought a temporary rise in living standards and escape from peculiarly southern forms of discrimination. As a result of war service, some blacks were able to encounter new cultures, to attend universities in Europe after the Armistice, and to be free for a while from the racial system of their homeland. Success against the enemy built pride. Despite attempts to restrict their training and keep them out of combat, many black soldiers learned to handle weapons, and some brought home arms, ammunition, and military skills that they put to use when whites attacked them in the postwar riots. This may have increased the white community's wrath and repression, but in at least one case it produced some constructive change. A historian who lived through the 1921 riot in Tulsa, Oklahoma, where blacks resisted white mobs, observed that afterwards racial justice became somewhat easier to secure.

On the whole, though, race conflict and disappointment led black Americans to feel that as far as their interests were concerned, the Great War had been fought in vain. After it ended, having received so little despite substantial cooperation and the sacrifice of young men in France, hundreds of thousands of blacks were so estranged from the United States that they thronged to a movement that called, among other things, for migration to Africa. A writer in *Challenge* magazine spoke for many blacks when he observed, after describing how an African-American child had been burned alive in Mississippi, "The 'German Hun' is beaten but the world is made no safer for Democracy."

The War and Women's Liberation: The Suffrage Campaign

One reason for the failure of blacks to gain from war service was that, for all its talk about democracy and freedom abroad, the Wilson administration had no intention of interfering with the racial order in the United States. But another was that African-Americans lacked the power and organization to induce their government to offer them significant concessions. In this respect, their situation contrasted with that of women suffragists who were entering the last phase of a long struggle that some of them considered fundamental to women's liberation.

Since the turn of the century, suffragists had built up state and local organizations patterned after the machines of regular political parties. With this machinery and using mass demonstrations and other techniques for promoting their cause, they won a series of victories, securing women the ballot and access to office in eleven states by 1917. Suffrage groups pressed Congress to pass a constitutional amendment that would give all American women the vote. They urged the president—who had said at various times that he had no opinion on the issue, that he could not speak out on suffrage until his party took a position on it, that suffrage was a matter for the states, or that he favored suffrage as a principle but did not favor a federal suffrage amendment—to give the amendment his public endorsement.

To the suffragists, as to other reformers, the Great War brought problems and opportunities. War work took time from suffrage agitation. The war gave arguments to their enemies who noted that women had played important parts in the anti-interventionist movement and that the only woman in Congress, Jeannette Rankin of Montana, had voted against the war resolution (along with fifty-five men). Anti-suffragists argued that women were incapable of taking part in a wartime government. One of their leaders, Henry Wise Wood, the president of the Aero Club of America, explained to the House Committee on Women Suffrage that America was in "the war of the brute," the type of war that only dominating men could handle. With the ship of state in wartime distress, there had to be "masculine men, not denatured men" on the bridge. Women were to play their part, Wood explained, "below decks."

Opportunities came in the form of arguments for suffrage. Suffragists observed how the country was addressing its wartime problems through national rather than simply local methods and contended that it was appropriate to give all the women of America the ballot through an amendment to the Constitution. Suffragists insisted that women deserved the vote for their wartime service to the nation. They had long demanded the vote as part of the traditional American right of self-government. Now they asked how America could continue to deny its own women that right when they were fighting for the self-government of other peoples. Before the war, many suffragists had claimed that women armed with the vote would reform society. They now contended that enfranchised women would help America reform the world.

Like other pressure groups, the advocates of woman suffrage had for many years appealed to prejudice and interest. They had solicited the approval of racists and nativists by insisting that white women and women born in the United States would outvote blacks and recent immigrants. During the war, suffrage leaders added new enemies to their list. "Every slacker has a vote," said suffrage leader Carrie Chapman Catt. "Every newly made citizen will have a vote. Every pro-German who can not be trusted with any kind of military or war service will have a vote. Every peace-at-any-price man, every conscientious objector, and even the alien enemy will have a vote. . . . It is a risk, a danger to a country like ours to send 1,000,000 men out of the country who are loyal and not replace those men by the loyal votes of the women they have left at home."

War affected the strategy and tactics of American suffragists as it affected their arguments. One group, the National American Woman Suffrage Association, had attempted for several years to win the vote state by state, gradually making the nation's politicians so dependent on women's votes that they would pass a federal suffrage amendment. The NAWSA supported candidates of any party who supported suffrage. Its tactics were sometimes aggressive but always respectable.

When the United States became a belligerent and the Wilson administration began mobilizing women, the NAWSA and its affiliates proclaimed their loyalty to administration policies. Many of the association's members felt the same surge of patriotism, the same joy in participation, that millions of other Americans experienced when their country went to war. Dr. Alice Hamilton noted a "strange spirit of exaltation among the men and women who thronged to Washington, engaged in all sorts of 'war work' and loving it." But to other women suffragists, who had been pacifists before the war, supporting the government's war activities was a matter of expediency.

Carrie Chapman Catt, who was the president of the NAWSA and who had been active in the peace movement before 1917, had little hope that Wilson's crusade would bring a better world. At the same time, she believed that antagonizing the government would not win women the vote. While insisting that the NAWSA concentrate above all on the suffrage campaign, she joined the Women's Committee of the Council of National Defense and encouraged the national association's rank and file to march in Liberty Bond parades, roll bandages, Americanize aliens, raise funds for the Red Cross and for military hospitals overseas, gather food-conservation pledges from housewives, and register women for war service.

Anti-suffragists chided the NAWSA for using the war to bargain with the government, for offering service in exchange for the vote—which, despite its denials, is what Catt's organization was doing, even if its members were acting spontaneously. In July 1917, an aide to Mrs. Catt told the president, "[O]ur hope has been to secure your interest and powerful influence . . . at the opening of the new Congress for a real drive for the enfranchis[e]ment of twenty million of American women, as a 'war measure' and to enable our women to throw, more fully and wholeheartedly, their entire energy into work for their country and for humanity, instead of for their own liberty and independence." The words and actions of the National American Woman Suffrage Association contained a double signal for the country and the government: grant women the franchise to reward them for patriotic service; give women the vote to ensure their full participation in the war.

Another group of suffragists, members of the National Woman's Party, pursued a radically different strategy. Influenced by the militant tactics of British suffragettes, the Woman's Party held the party in power responsible for the fate of the suffrage amendment. Since the amendment had failed to pass while the Democrats controlled Congress and the presidency, the Woman's Party tried to drive the Democrats out of office in the election of 1916. When that did not work, it attempted to harass and humiliate the president into pushing their amendment through the legislature.

In January 1917 Woman's Party pickets began to assemble outside the White House, and for the next year and a half, over a thousand of them—some of them workers, several of them recent college graduates, many of them members of well-known, privileged families—marched silently outside the gates to the presidential mansion. Their banners asked, "How Long Must Women Wait for Liberty?," described the president as "Kaiser Wilson," and reproached him with his own words:

WE SHALL FIGHT FOR THE THINGS WE HAVE ALWAYS HELD NEAREST OUR HEARTS—FOR DEMOCRACY, FOR THE RIGHT OF THOSE WHO SUBMIT TO AUTHORITY TO HAVE A VOICE IN THEIR OWN GOVERNMENTS

The administration and anti-suffrage congressmen became increasingly angry. To Wilson, the Woman's Party seemed "bent on making their cause as obnoxious as possible." A senator denounced the party for "selfish and silly disregard" of the need for national unity. Still, the militants kept returning to the picket line. "If a creditor stands in front of a man's house all day long," declared Alice Paul, one of the party's leaders, "demanding the amount of the bill, the man must either remove the creditor or pay the bill."

Attempts were made to remove the creditor. Sailors, marines, and other men began to attack the pickets, grabbing and destroying their banners. The police arrested the marchers—for unlawful assembly and obstructing the sidewalk. Sentenced to the workhouse for up to seven months, party members insisted on being treated as political prisoners. They defied prison discipline and refused to eat. The authorities threatened them and treated them roughly. Guards handcuffed women who refused to cooperate, tied some to the prison bars, twisted their arms behind them, hurled them to the floor, and held them down while prison doctors forced food into their stomachs down tubes pushed through the nose or mouth. Alice Paul spent over a month in solitary confinement and a week in a ward for psychopaths. The jailers' methods only strengthened the resolution of the Woman's Party, which printed smuggled-out stories about the treatment of the prisoners.

The party's critics, including other suffragists, claimed that these tactics weakened the cause, depriving politicians of a chance to accede to woman suffrage gracefully. The inmates' activities certainly upset President Wilson. He called in the Washington police commissioner and told him that a terrible mistake had been made. Wilson tried to pardon the pickets, but they refused to accept. Eventually, they agreed to go free, then returned to the White House, resumed picketing, set fire to speeches by the president that contained the word "democracy," and burned him in effigy.

Finally, it was the government that gave way. "My God," said a New York politician, "we'd better do something to satisfy those hellions." That is what Congress did, encouraged by the president, who let it be known that he was responding to the polite requests of Mrs. Catt's organization. The House passed the suffrage amendment on January 10, 1918, a year to the day from the first appearance of members of the Woman's Party outside the White House. A few weeks before the congressional election of 1918, Wilson publicly endorsed the amendment. "We have made partners of the women in this war," he said. "Shall we admit them only to a partnership of suffering and sacrifice and toil and not to a partnership of privilege and right?"

While the federal suffrage amendment did not make it through Congress until the next year and was not ratified until 1920, the war clearly influenced its passage. All parties to the struggle over woman suffrage sought to make the war serve their own ends. Anti-suffragists tried to use it to show that women were unqualified for the ballot. Moderate suffragists pointed to their pro-war activities as evidence that women were fit to vote and discreetly threatened to limit what they did for the war if the government did not respond to their campaign. Militant suffragists embarrassed the administration and attacked its credibility by contrasting the official war aims with the way the government treated women. President Wilson employed the war to justify changing his position on suffrage. When he claimed that the vote was a reward for wartime loyalty, he manipulated the suffrage campaign to increase the support of women for the war.

World War I, then, advanced the movement for women's votes that had acquired substantial momentum by the time America became a belligerent. While suffrage was only one part of the struggle for women's liberation, whose fulfillment was still far off when the Great War ended, war did advance equality for women in other ways. It opened economic opportunities for women, enabling them to take jobs in government and private industry and making it possible for some to gain promotions and higher pay.

These were limited and temporary achievements. Even during the war, female employees were often confined to "women's jobs," earned less than men for similar work, and were subject to sexual harassment. Most of the women who filled wartime openings had probably been working before the war began, and when it ended, many of them, particularly those who had begun to perform "men's work," lost their positions. In New York City, for example, twenty women judges were forced to resign immediately after the Armistice, and women who worked on railroads during the war were replaced by returning servicemen, ousted through the application of laws designed to protect women from hazardous conditions or just summarily removed. Still, war enabled thousands of women to widen their experience and enjoy a taste of independence, at least temporarily. And it became the occasion for an effort by the United States government to advance campaigns in which millions of American women were participating—against alcohol abuse and for the abolition of a double sexual standard.

7 ‖ The Great War, Prohibition, and the Campaign for Social Purity

Several years before the war, proponents of temperance and abstinence had begun a national campaign to restrict alcohol consumption. Challenged by an expanding liquor industry and the multiplication of saloons, alarmed by the havoc that drunkenness wreaked on family and society and on individual health and morals, they had formed an Anti-Saloon League and a Women's Christian Temperance Union. Prohibitionists developed informal alliances with suffragists and Protestant churches, fielded an army of speakers, created elaborate political machines, and won numerous victories in towns and counties, cities and states. After the election of 1916, there were nineteen dry states and majorities in both houses of Congress that favored restrictions on the sale of alcoholic beverages. Backed by members of the medical profession who considered alcohol a poison; by social scientists who linked alcohol consumption with poverty and political corruption, crime, and prostitution; by thousands of businessmen who regarded drinking as a menace to productivity; and by women advocates of temperance who now could vote in several western states, the organized opponents of drink had become one of the most formidable pressure groups in America.

Like other pressure groups, prohibitionists tried to turn the war to their advantage. They spread the idea that beer consumers were disloyal and pro-German, a theory to which beer manufacturers clumsily contributed. Never known for political finesse, brewers had given money to a German-American Alliance in the belief that the alliance would fight prohibition. During the spring of 1916, the New York *World* published documents purporting to show that the alliance had participated in pro-German activities. Congress revoked its charter, and the United States attorney general accused the brewing industry of unpatriotic activity. Prohibitionists publicized these stories, warning America that "pro-

Germanism is only the froth from the German beer saloon. . . . Kaiser Kultur was raised on beer [and] sobriety is the bomb that will blow Kaiserism to kingdom come."

Prohibitionists found other arguments in the war. They had always talked about the way alcohol abuse disrupted families and damaged children. Now they demanded that beer and liquor makers stop using grains while children in the warring countries were going hungry. To overcome the resistance to a federal prohibition amendment, particularly strong among advocates of states' rights in the South, they noted, as other reformers did, that war was making federal action acceptable to the nation. Like other reformers of the time, they claimed that their particular program would increase wartime efficiency.

The reputed connection between efficiency and abstinence had been talked about for years. On the principle that sober workmen were more efficient than those who drank, several businessmen had come out for prohibition before the war. Social-welfare reformers favored it as a way of helping poor people function efficiently in their struggle against an oppressive environment. Now sober efficiency would vanquish America's foreign enemies. "Drastic prohibition would increase the efficiency of the Army," declared an editor of *Manufacturer's Record*. "It would increase the efficiency of the workers in industrial plants and mining operations. It would increase the efficiency of men on the farms and add enormously to the potential power of the nation." Commenting on a proposal to limit production of intoxicants, Senator Kenneth McKellar of Tennessee remarked, "I believe it would make us more efficient as a nation in the time of this emergency . . . if the liquor traffic is abolished. . . . No drunken man was ever efficient in civil or military life."

The war encouraged the prohibition movement by stirring idealism, heightening the national sense of mission, and encouraging a cult of purification—the cleansing of physical and social poison from mind and body, community and nation. Reverend C. C. Bone was no more idealistic than millions of his countrymen when he exulted to a cheering Anti-Saloon League audience: "America will 'go over the top' in humanity's greatest battle and plant the victorious white standard of Prohibition upon the nation's loftiest eminence. . . . With America leading the way, with faith in Omnipotent God, and bearing with patriotic hands our stainless flag, the emblem of civic purity, we will soon . . . bestow upon mankind the priceless gift of World Prohibition."

Yet opponents of prohibition also found the Great Crusade a source of arguments. On behalf of workers in the industries that made and sold alcoholic beverages, Samuel Gompers warned the president that if Congress passed a prohibition amendment it would produce discord among the people and divert their minds from the winning of the war. Perhaps two million workers would lose their jobs, leading them to become "disaffected." A distillers' lobbyist wrote members of Congress that alcohol should be provided to soldiers to steady their nerves and to relieve them of the terror of advancing against machine guns. Senator James Phelan of California, sensitive to the views of California vintners, told his colleagues that claret gave the French endurance, courage, and élan. And James Wadsworth of New York reminded the Senate how badly war workers needed cheerfulness and how unjust it was to tell them they must labor to the limit of endurance and then say, "You shall not have a glass of beer."

Although President Wilson preferred not to become entangled in the conflict over prohibition, his administration and Congress undertook a series of war measures that turned the nation dry. Congress, which had already stopped liquor sales on army posts and at old-soldiers' homes before the war, provided in the Selective Service Act of 1917 for dry areas in and around military camps and forbade people to sell or give drinks to servicemen, even in private homes. The American government told the French not to sell doughboys anything stronger than wine or beer. In August 1917 a section of the Lever Act outlawed the use of foodstuffs for distilling liquor, and in early December the president prohibited the brewing of beer with more than 2.75 percent alcohol and severely restricted the amount of grain that brewers could use. On December 22, 1917, Congress continued the wartime campaign against alcohol into the indefinite future when it approved the Eighteenth Amendment, which was ratified two years later. At midnight on January 16, 1920, all remaining legal saloons and barrooms closed.

Thus began one of the most ambitious attempts ever made to reform American culture, to purge the nation of a substance that prohibitionists believed had ruined the health, weakened the morals, disrupted the society, and corrupted the politics of America. The United States government responded favorably to the political strength of the prohibitionists because of the personal sympathy certain government officials had for their cause and also because prohibition was a way of managing the nation at war; it held promise for diverting food supplies to war production, increasing the output of war workers, and building a fighting force that could be sent into battle physically and morally strong and soberly efficient.

Social Purity and the Great Crusade

The struggle against alcohol abuse had long been intertwined with a campaign for social purity, an effort to wipe out prostitution. The enemies of prostitution regarded it, as the prohibitionists regarded alcohol, as a source of moral degradation, social disorder, disruption of family life, and physical illness. During the war, the social purifiers, like the enemies of drink, presented their program as essential to sustaining the military effectiveness of American troops. They viewed the struggle against the brothel as a way of preserving the dignity and welfare of women while guarding American society against the disease and corruption that war had brought to other nations.

In America, the social-purification campaign began in the nineteenth century, starting as a crusade against the country's red-light districts. Social purifiers, many of them former antislavery workers or descendants of abolitionists, proposed to treat prostitution as they or their ancestors had treated slavery. They did not wish to contain it; they intended to get rid of it completely.

At first, the purifiers failed to make any serious impact on public policy. Aside from the power of brothel owners, they had to combat a structure of beliefs that included the notion that males should not be restricted to monogamy and the view that organized prostitution prevented the seduction of respectable women. But in the years before World War I, their campaign benefited from the hopeful spirit and from the methods of progressivism. Muckraking journalists who had sensationalized political corruption and the rise of trusts began to investigate the white slave traffic. Such reformers as Jane Addams explained how urban slum conditions forced women to become prostitutes. Feminists linked the struggle against vice with their campaign for women's equality. In 1914 social purifiers organized the American Social Hygiene Association to educate people about vice and to lobby for its eradication. A few big-city governments shut down red-light districts. Meanwhile, the purifiers began to attack prostitution in the army.

The United States Army was a conspicuous target, for it had traditionally accepted prostitution as a necessary outlet for its men and tried to control it by licensing prostitutes and inspecting them for venereal disease. During the War for Filipino Independence, it had run the largest licensed brothel in the world. This policy of regulated vice, never acceptable to social purifiers, became intolerable as the army suddenly grew from a small Indian-fighting and colonial force into a vast organization with posts all over the country.

In 1916 an incident occurred that drew national attention to the issue of military vice. During the Mexican border mobilization, rows of saloons and brothels sprang up on the perimeters of army camps, and National Guardsmen swarmed through red-light districts in nearby towns. Questions raised by these events lingered into the following year when the AEF began to form. If these citizen-soldiers had succumbed so easily to the temptations of the border towns, what would happen when innocent American troops reached France, a country that accepted and regulated prostitution? Would doughboys contract venereal disease *en masse?* Was the United States to send millions of impressionable young men to save Europe from the German autocracy and then bring the survivors home corrupted and sick, an unmanageable threat to the health and morals of the American nation? A soldier's mother, Mrs. E. L. Wunder, stated the moral issue to the president: "We are praying you will make camp life clean for boys [who] are willing to offer their bodies as sacrifice for their country, but not their souls."

The men Wilson appointed to head the armed forces shared Mrs. Wunder's concern. Under the leadership of Newton D. Baker and Josephus Daniels, the War and Navy departments became schools for social reform and personal improvement and sought to uplift America's soldiers and sailors, to indoctrinate them in such middle-class customs as frequent bathing, to regulate their drinking habits, to inculcate them with principles of good government, and to make them literate. Under the leadership of these reformers, the armed services developed techniques for keeping American youth healthy and pure, methods aimed at ensuring the efficiency of the U.S. military machine.

Secretary Baker outlined the War Department's social purification plan in a 1917 speech to recreation specialists. The boys leaving for France, Baker said, would

> face conditions that we do not like to talk about, that we do not like to think about. They are going into a heroic enterprise and heroic enterprises involve sacrifices. I want them armed; I want them adequately armed and clothed by their government; but I want them to have invisible armor to take with them. I want them to have an armor made of a set of social habits and a state of mind . . . so that when they get overseas and are removed from the reach of our comforting and restraining and helpful hand, they will have gotten such a state of habits as will constitute a moral and intellectual armor for their protection overseas.

As American white servicemen tried to sustain their native system of race relations by bringing Jim Crow across the Atlantic, so the War Department attempted to maintain the American moral code,or what social purifiers wanted the American moral code to be, by exporting it to France.

How to forge the invisible armor? The first step was to wipe out prostitution around military bases in the United States or, if that was impossible, to force it out of sight. For this work, the War Department employed a Committee on Training Camp Activities, staffed by prominent social reformers under the leadership of Raymond B. Fosdick.* At thirty-three, Fosdick was already experienced in social work and city government and had made himself an expert on police systems. He had become an authority on military vice during an investigation of conditions at the Mexican border.

Following the social purification approach, the Committee on Training Camp Activities tried to abolish red-light districts. It assumed that physical examinations could not guarantee the health of prostitutes and that prostitutes would not restrict themselves to designated areas. The committee also felt that red-light districts offered too great a lure. Since many people, including some public officials, did not share these assumptions or sympathize with the goals of Fosdick's group, the CTCA ran into problems. One of its investigators reported that the citizens of New Orleans seemed to feel that there was nothing wrong with prostitution. Prostitutes expelled from brothels there moved to new quarters and sometimes turned up again at their old stations. The army had to declare Birmingham and Seattle off-limits. Philadelphia was reformed only after marines were sent to patrol the streets.

Nevertheless, according to Fosdick, every red-light district in America was closed by the end of 1917—by Fosdick's count 110 in all. Even if we allow for inaccurate reporting, this was surely one of the Progressive movement's most remarkable accomplishments.

The next step in constructing the invisible armor was to dissuade recruits from consorting with prostitutes. The War Department sent draftees a *Home Reading Course* that exhorted them not to take a last night's fling that might land them in the hospital or make them too ill to take up their new duties "with profit and enjoyment." It urged these citizen-soldiers to convey the impression that they were "too much in earnest in the great task assigned to them to indulge in lewdness and vice." Live a clean life, it told them. Avoid illicit intercourse and shun those things that tended to promote sexual excitement and desire, particularly "obscene conversation, reading matter and pictures."

The CTCA's division of social-hygiene instruction could not have been more blunt in its warnings to the troops. It turned out lurid films on keeping "fit to fight" and issued pamphlets that asked such questions as, "You wouldn't use another fellow's toothbrush. Why use his whore?" and, "How could you look the flag in the face if you were dirty with gonorrhea?" Its posters, plastered on army-camp buildings, warned that "A German bullet is cleaner than a whore." One lecturer told the troops to

*Fosdick also headed a navy training-camp committee whose operations paralleled those of the CTCA.

"live strong and clean" and "save every drop of your strength and man-hood for the supreme experience." The CTCA tried to prevent soldiers from even thinking about sex. One of its lectures urged them to forget about their sex organs. "A man who is thinking below the belt," an official CTCA syllabus declared, "is not efficient."

The committee cautioned its speakers not to alienate their audience with moralizing and sentimentality and urged them to offer "man-to-man talk, straight from the shoulder without gloves." However, the pamphlets, stereopticons, movies, and the photographs of open lesions and physical deformities resulting from venereal disease that constituted the CTCA "educational prophylaxis" program seem to have produced another un-wanted effect. Its exhibits were so horrifying that an army doctor thought they conditioned soldiers to associate all sexual relations with disease. One of the lecturers told Fosdick what it was like to impart the CTCA message:

> As I looked into that mass of faces my courage grew . . . to a heated pitch. I stripped the moral issue to the naked bone. . . . I defied them. I schooled them with a world of medical facts . . . on the ravages of venereal dis-ease. . . . I sneered at them for lacking the moral courage to fight the common tendency in armies. . . . I appealed to their regimental pride. Before long they were virtually as pliable as putty. . . . IT WAS THEN THAT I DELIVERED THE GOVERNMENT MESSAGE. . . . I told them that they had to be 100% efficient to win this war. . . .

The third step that was taken toward building the invisible armor was to create a moral equivalent for sex, an action exemplifying the indirect approach favored by managers of the American war effort. In his speech to the recreation specialists, Secretary Baker recalled how urban reformers had found that restrictive yet ineffective laws could not overcome the temptations of a great city. They had discovered that the best way to deal with the problem was to offer adequate opportunities for "wholesome recreation and enjoyment." Raymond Fosdick observed, "It does not do any good in dealing with red-blooded young men merely to erect a series of *verboten* signs along the roadside; you must have something positive for them to do." Applying this principle to the army, the CTCA, in conjunc-tion with the Young Men's and Women's Christian associations, other religious agencies, the American Civil Liberty Association, the War Camp Community Service, and local groups, set up such wholesome activities for the troops as athletic programs, community socials, and group singing. YMCA-sponsored "hostess-houses," where soldiers could meet female guests in quiet homelike surroundings, were especially popular with young men who, as Baker put it, "spontaneously" preferred to be "decent."

Despite their success in eliminating organized prostitution and careful as they were to divert sexual impulses into sport, reading, and "clean fun," government officials feared their program would fail if ordinary young women succumbed to "the lure of the uniform." The source of VD in most of the camps, as Fosdick explained, was no longer the prostitute but "the type known in the military as the flapper—that is, the young girls who are diseased and promiscuous." To seal this opening in the invisible armor, to protect the troops against what a CTCA official called "women who would destroy their fighting power," the War Department established a Committee on the Protection of Girls under the direction of Maude E. Miner, secretary of the New York Probation and Protective Association.

This agency sent a hundred fifty field workers to the vicinity of camps to uncover and eliminate illicit activities. It urged communities to close burlesque shows and other temptations that might stimulate soldiers and their women friends. Its agents followed soldiers and their lovers into secluded places and pulled out women who were hiding in trenches. Miner's committee warned young women not to arouse impulses in the troops "which they cannot express without weakening the army" and told them they had no right to "make the sex impression upon any one of these soldiers, even in the most refined, indirect unconscious way—through dress, position in dancing, slight familiarities, suggestive looks or innocent flirtation." The American soldier, one official observed, must find American girls "such inspiring friends that he will be proud to risk his life for . . . American womanhood."

As additional precautions, local and federal authorities forced women in the vicinity of camps to undergo compulsory physical examinations and staged vice raids that swept up mostly unemployed, working-class, and unescorted women, who were denied bail while they were tested and treated. The president and Congress allocated hundreds of thousands of dollars to establish federal detention centers to which over eighteen thousand women and girls convicted of prostitution were committed from 1918 through 1920.

Though the War Department went to great lengths to insulate American doughboys against vice, it was ready in case moral purification failed. The army made prophylactic instruction and treatment available to all troops. Secretary of the Navy Daniels, however, forbade the use of prophylactic kits on the ground that using it "would tend to subvert and destroy the very foundations of our moral and Christian beliefs and teachings with regard to these sexual matters." But when Daniels was out of his office, at Fosdick's suggestion Assistant Secretary of the Navy Franklin D. Roosevelt signed an order providing for prophylactic treatment.

As Secretary Baker had anticipated, the supreme test of the invisible

armor occurred in France. The French government disagreed with the
social purification approach. While the French wished to protect their own
population from venereal disease, they did not feel that closing brothels
was the way to do it. Indeed, they feared that if troops could not have sex
with prostitutes, they would seek an outlet among respectable women.
French officials doubted that anyone could sublimate the sexual drives of
an army, and some of them felt it was wrong to meddle in affairs of sex,
for, as one of them remarked, if a man and a woman wanted to sleep
together, it was no one's business and no one else should interfere. Accordingly, Premier Clemenceau offered to have his government construct
brothels for the AEF, and the foundations were laid for three. When
Fosdick reported the plan to the secretary of war, Baker exclaimed, "For
God's sake Raymond, don't show this to the President or he'll stop
the war!"

The question of what to do with existing brothels remained a source of
friction between the social purifiers in the American government and
French municipal authorities. Local French officials prevented the Americans from closing the whorehouses of Saint-Nazaire. The army then declared them off-limits and stationed MPs at the doors. The U.S. provost
marshal proclaimed the brothels of Blois out of bounds when American
troops arrived, but following French protests and the recommendation of
an American medical officer, they were reopened. Once, some local U.S.
Army authorities reverted to the traditional method of their service and set
up a brothel with separate visiting hours for officers and enlisted men.
After it had been operating for six weeks, AEF General Headquarters
declared it off-limits.

Most of the time, the American forces in France followed the social-
purification approach. General Pershing, who had instituted legalized
prostitution on the Mexican border and had himself suffered from gonorrhea twice, was determined to make the AEF physically and morally
"clean." He cabled the War Department to withhold part of the soldiers'
pay so that they could not afford prostitutes. On July 2, 1917, he issued a
general order making contraction of venereal disease a court-martial of-
fence. Thirteen months later, Pershing, who visited his own lover every
time he went to Paris, issued a bulletin declaring that "sexual continence is
the plain duty of members of the AEF." To limit opportunities for meeting
French women, the army strictly limited the free hours of enlisted men in
and around Paris, while the YMCA conducted sightseeing tours of the
provinces and presented wholesome, diverting entertainment.

How effective was the army's social-purification approach? This question really has two parts: How well did the army keep venereal disease

from reducing its fighting power, and how well did it preserve the moral system approved by men like Fosdick and Baker? Official figures show that the army was quite successful in keeping venereal-disease rates down, though it achieved results that varied with different classes of soldiers. For the year ending August 30, 1918, the average rate was 126 cases of VD for every 1000 men. The army medical department believed that five out of six cases had been contracted before the victim entered the service. In France, the officers' rate rose to 27.7 cases for every 1,000, triple the level in stateside army camps. But for enlisted men, particularly those in the trenches, the rate dropped sharply.° As a result, one objective of the social purifiers was achieved. Although eighteen thousand of the two million AEF troops were out of action with venereal disease on an average day (enough to fill several regiments), VD did not seriously undermine the American combat forces.

It is more difficult to determine whether American soldiers behaved as the purifiers wished. Venereal-disease rates are inconclusive in this respect, for they include tens of thousands of men who spent so much time in the trenches that they rarely saw a woman in their sojourn in France. Furthermore, the rates give only the incidence of reported disease and not of sexual activity. Since the army taught its men how to avoid disease while having sex and since the AEF punished soldiers who reported ill with VD, doughboys could have slept with French women without ever registering as venereal-disease patients.

Fragmentary reports about the moral behavior of American troops point in more than one direction. A survey of World War I veterans conducted in the 1970s suggests that the army's anti-vice campaign made a considerable impact but fell short of its objectives. To a question about the extent to which troops had "consorted" with women, Corporal Edmund A. Grossman of the Thirty-fifth Division responded that there had not been a lot of consorting because "we were afraid of disease, which some men got." Harvey C. Maness, a private in the Thirtieth Division, remembered that his buddies had just flirted as a rule, though there were occasional contacts with "improper" women. "We had been admonished to be gentlemen," he wrote, and "drilled" to understand the tragedy that venereal disease could bring. Yet Arthur E. Yensen, a wagoner in the Fifth Division recalled that the way doughboys treated European civilians, "judging all their women by the whores who were too easy to meet," had made him ashamed to be an American.

°MPs, who guarded whorehouses, had the highest rate in the army.

Sergeant Merritt D. Cutler, who served with the Twenty-seventh Division, told the oral historian Henry Berry:

> [O]ne thing I definitely didn't do was go into one of those crummy whorehouses they had. Those lectures they'd given to us in Spartanburg had really gotten to me. You know, all those pictures they showed us of men with their noses falling off—yes, and other things also.

But this same veteran reported passing a dirty-looking bordello near the Belgian border and seeing a line of French, British, and American soldiers about a quarter of a mile long waiting "for a quick roll in the hay." Sergeant Cutler imagined that the place must have smelled terribly. "No sir," he said, "sex on the assembly line wasn't for me."

Other information suggests that Pershing's Crusaders, and not just the ones Cutler saw on that line, acted like armies that had never been clothed with an "invisible armor." In almost all AEF units, over half the soldiers requested prophylactic treatments at least once, and at Saint-Nazaire during one six-month period, a few men were treated thirty or forty times each. In the leave area at Nîmes, 3,712 out of 3,934 men applied for prophylaxis. An article in the *Journal of the American Medical Association* reported that "many soldiers, even though thoroughly instructed in venereal disease matters and fully provided with the opportunity of YMCA and Red Cross entertainments, were obstinate in preferring the society of women of the Paris streets." A Frenchwoman complained to the army that an MP had followed her closely and insolently. "I sometimes ask myself," she wrote, "what American education can be that it allows of such disrespect of women. It must not be believed that in France, all are prostitutes and that women are in the world simply to amuse your soldiers. . . ." Frank Tannenbaum wrote from an army camp in the spring of 1919 that prostitution still went on. His fellow enlisted men still craved sex, he said, purely physical sex. Tannenbaum thought their morals were much lower than before they entered military service. Colonel George Walker, the venereal-disease expert, sent investigators to cafés to interview prostitutes, who almost invariably reported that AEF troops had come to prefer oral sex—"the French Way." Walker remarked,

> When one thinks of the hundreds and hundreds of thousands of young men who have returned to the United States with those new and degenerate ideas sapping their self-respect and thereby lessening their powers of moral resistance, one indeed is justified in becoming alarmed.

One can only conjecture whether the sexual activities of doughboys would have been substantially different if Fosdick and Baker had never tried to give them an invisible armor. There were no Kinsey-type surveys to describe their sexual behavior before and after they entered the armed forces and only scattered information (for instance, the reported existence of 110 red-light districts) about the morals of Americans in general. Possibly the army encouraged greater frankness about sex by making it the subject of public lectures and training films; perhaps it made some men who feared syphilis and gonorrhea more willing to seek out sexual contacts because the army had taught them how to have intercourse without contracting venereal disease. It may be that the crackdown on professional prostitution helped change the morals of American women by clearing the field for amateurs.

Still, it can be said that the *attempt* to regulate the morals of the AEF was a classic instance of the way the United States government tried to manage its people in wartime. Using an indirect approach, offering substitute gratifications more than punishment, fiercely regulating those with little power—women who were accused of prostitution—groups of government officials sought to control and divert the tides of libido in a multimillion-man army. Responding to a movement to end the degradation of women and acting on behalf of military efficiency and of the health, welfare, and moral purity of the nation, they tried to break down an ancient tradition of sexuality, to manipulate drives and emotions, and to encourage young Americans to make war instead of love. Their optimism was remarkable; their faith in reason's power over instinct, awe-inspiring.

It should now be apparent that in these years a great world war was very compatible with particular elements of the national reform movement. War helped the advocates of prohibition advance their program. It assisted women suffragists and may have "uplifted" the morals of some doughboys. It brought benefits of a welfare state to millions of servicemen and their dependents and to workers in war industry. American experience in the Great War showed that a bloody conflict overseas, at least a short one, does not necessarily make social reform impossible at home.

Yet the war did not benefit all reform movements equally. The unevenness of its effects resulted partly from contradictions within the attitudes and programs of reformers. Progressives who objected to centralizing the management of society could not be as pleased with the war welfare state as those who favored national planning and regulation. Insofar as wartime events satisfied reformers who wanted racial or sexual equality, it dissatisfied progressives who preferred to keep women and

blacks in their places. To reformers who resented Victorian moral restrictions, the wartime campaign to regulate morals seemed misguided at
best.

Another reason for the unevenness of reform lay in the government's
method of responding to interest groups, its way of distributing rewards
according to the power of those groups and their usefulness to its purposes. By the time the United States entered the war, the suffrage movement had developed effective political machines and an apparatus for
generating publicity, both of which could be used to help the administration mobilize women or to disrupt its efforts. Prohibitionists were numerous and well organized. Workers, especially those with skills desperately
needed in factories and shipyards, had considerable power, including the
ability to disrupt war production. Soldiers and their millions of dependents
constituted a potential source of political influence unparalleled since the
days of the Grand Army of the Republic. Blacks, although numerous,
could not vote in many states and, having neither power nor highly developed political organizations, were able to extract little from Congress or
the administration.

The Great Crusade marked one of the high points of social reform in
the twentieth century. Both idealism and the progressive aims of members
of Wilson's administration had much to do with this fact; so did a desire to
prevent the war from weakening American society. But World War I also
promoted reform because reformers had the resources and the will to
bargain with their government as it sought to manage the nation.

8 ||| American Intellectuals and the Control of War: Dewey, Lippmann, and Bourne

Intellectuals were as divided over their nation's entry into war as the rest of the American people, many welcoming it, others abhorring it or facing it with resignation. Like businessmen and labor leaders, suffragists and moral reformers, certain intellectuals tried to use war to promote causes and interests. Some saw it as a means to advance their careers, to make their lives more exciting, or to demonstrate to their fellow citizens the value of people who dealt in ideas. One audacious group of liberal intellectuals, including John Dewey and Walter Lippmann, believed they could control the war, determine their government's war aims and the terms of the peace settlement, and use the war to reshape the United States and other nations along liberal lines. To these optimistic notions, the radical intellectual Randolph Bourne demurred. In disputing whether and how the war could be controlled, these three men performed traditional functions of intellectuals: Bourne as alienated critic of the established order; Dewey and Lippmann as rationalizers for those in power. The interesting question is why in this case they assumed their respective roles.

John Dewey and the War as an Experiment

John Dewey, born in 1859, had sprung from generations of Vermonters, serious people, pious and thrifty. His mother was a devout evangelical Protestant whose prosperous farm family adhered to a tradition of public service, sending some members to the Vermont General Assembly and the United States Congress. His father started out as a farmer, then opened a grocery in Burlington, and later went into the cigar and tobacco business. John Dewey attended the University of Vermont, where, in his senior year, philosophy captivated him, and he decided to make it his profession. After a few years teaching high school, he entered the Johns Hopkins University, where one of his fellow graduate students was Woodrow Wilson, and earned a doctoral degree. He spent some time in midwestern universities—Michigan and Wisconsin—developed his reputation, and taught from 1894 to 1904 at the University of Chicago with such illustrious colleagues as the economist Thorstein Veblen. Then he left for Columbia University, where he served on the faculty of the philosophy department and in Teachers College.

Dewey looked the part of an academic philosopher, with baggy suits, slightly disheveled hair, small drooping mustache, and tie a bit askew. A mild-seeming man, he peered at his class through rimless glasses, turning now and then to gaze into the distance as he talked. His seemingly distracted, apparently rambling manner put off some students; but the discerning realized that he was thinking while he spoke, and they appreciated his wide-ranging intellect and sweet disposition. These students did not just admire Dewey; they loved him. They understood that he was one of America's most eminent philosophers, a founder, along with William James of Harvard, of the philosophy known as pragmatism or instrumentalism.

Like numerous contemporaries, Dewey was deeply affected by the picture of the world Charles Darwin had drawn in *The Origin of Species*—a view of organisms confronting a dynamic environment in which organisms best able to adjust to their changing surroundings were most likely to flourish and the kinds that failed to adapt tended to die off. Dewey regarded human beings as a special variety of organism that not only adjusted to the environment but also consciously shaped it by using a special quality of the human species—intelligence.

To John Dewey, the highest form of intelligence was the scientist's method of producing hypotheses and experimenting to see if they worked. When facing a problem whose solution was not clear, the intelligent person thought about alternate courses of action, developed ideas about what to do, and imagined how a particular plan of action would work in practice. Then by carrying out the most promising of these plans, the thinker determined which ideas were true—that is, which ones solved the problem.

Dewey recognized that people did not always apply intelligence to improve their circumstances. He knew that science had sometimes been used to harm society. He had seen how unregulated capitalism employed science and technology in ways that made the environment inhospitable to humans, producing social stress, alienation, and fragmented communities, enriching an elite while permitting masses of people to suffer. He believed society found it hard to cope with modern conditions because people did not change their ideas, values, and institutions rapidly enough to keep pace with scientific advances.

Dewey proposed to use philosophy to clarify issues that underlay social conflict, to promote collective solutions to social problems, and to foster a "public control" that would redirect industry into socially beneficial channels. On the principle that when the world was in flux it was more susceptible to human management than when conditions had solidified, he welcomed disorder. The philosopher's task, he maintained, was to keep thought from ossifying and to help people reshape a malleable society by applying intelligence, self-conscious thought directed at adjusting human beings to their environment. He hoped to reform American education so that instead of acquiring the obsolescent facts and fossilized individualistic values of their elders, children would develop their natural tendencies to cooperate, discovering in the process how to serve and adjust to a changing society, to make it "worthy, lovely, and harmonious." He believed that even a catastrophe like World War I could be used to foster constructive social evolution.

Yet the philosopher was no war lover. In 1908 he warned that military preparedness could tempt nations to use their weapons irresponsibly, and he challenged the idea of militant Darwinists that war kept the race from degenerating. He sympathized with organizations like the Woman's Peace party that wanted to stop World War I and keep America out of it. Nevertheless, in 1916 he backed American rearmament and after the United States entered the struggle served as a consultant to the State Department, worked for the Columbia University Division of Intelligence and Publicity, and published a series of articles in the *New Republic* and other liberal journals, supporting American intervention.

Though Dewey was far from delighted when the United States went to war, he saw opportunities in that event. While others complained about the chaos war created, Dewey believed that America's entry had created a "plastic juncture," a chance to reorganize the world economy and to establish a "public control" across national boundaries that could solve social problems. He thought the war would increase the use of science for communal purposes and believed American participation would help produce an international body that would adjust conflicting interests of nations, preventing future war.

The important thing was to ensure that the war was actually fought for those objectives and that the Allies and the United States used force in a rational, efficient manner.* He wanted the Wilson administration to bargain with the Allies to ensure that their war aims conformed to American ideals and regretted that it "romantically abstained" from doing so. Still,

---

*Dewey distinguished force from violence, which he defined as force used in an unintelligent, wasteful manner.

he hoped the American people would compel the Allies to accept the objectives their president had proclaimed.

Dewey understood that when fighting a big war, the government was bound to manipulate public opinion. "Practical political psychology," he noted, "consists largely in the . . . expert manipulation of men *en masse* for ends not clearly seen by them." While this was a threat to free expression, freedom of speech had survived worse crises, he said, and would survive this one. At the same time, he warned that extreme efforts at thought control would not work and urged the Wilson administration not to appeal to hatred and fanaticism but to allow Americans, even in the midst of war, to debate the war aims and the ways to secure them. He recommended that it take the public into its confidence and, in a businesslike manner befitting a businesslike nation, discuss with the American people ways of securing its goals. The people would then wage the war, not as hypnotized fanatics but as thinking men and women.

Events disappointed him. As he watched fanaticism grow and saw the government crush freedom of speech, he decided that the only place where one found comparable repression was Germany. The national and international reconstruction for which free expression was to be temporarily suspended never took place. He blamed this on the American people and the United States government. The peace settlement failed, he said, because America had entered the war idealistically, providing the Allies with huge quantities of aid without insisting that they lay out their real objectives, and because

> we took into the war our sentimentalism, our attachment to moral sentiments as efficacious power, our pious optimism as to the inevitable victory of the "right," our childish belief that physical energy can do the work that only intelligence can do, our evangelical hypocrisy that morals and "ideals" have a self-propelling and self-executing capacity.

Though he conceded after the war that the pacifists whom he once had chastised now seemed much righter in retrospect, he did not consider, or at least did not say, that he had courted disillusion by imagining the Great War could be made to serve his liberal ends.

Walter Lippmann and the House Liberals

The publicist-intellectual Walter Lippmann followed a trajectory closely parallel to Dewey's. Lippmann was born in New York City toward the end of the nineteenth century to prosperous German-Jewish parents who indulged and protected him and sent him to the best schools and took him to absorb high culture in Europe. Well-dressed and well-mannered, a good, obedient son, Lippmann in his twenties always wore his rubbers when it rained.

He was a prodigy, brilliant at writing, a fluent speaker, and admired by fellow students for his intellectual abilities, though some thought him slightly aloof and a little old for his age. Others viewed him as an outsider. At Harvard, snobbery and anti-Semitism hurt him, though he concealed their effects. His family had tried to assimilate to the dominant American culture, and he did the same, identifying with Anglo-Saxon Protestants, later entering their clubs and governing boards as a token Jew, attaching himself to members of the establishment, and using intellect and social skills to make himself a man with influence.

Lippmann was quick to absorb the notions of advanced philosophers and scientists—Nietzsche, Graham Wallas, Freud, Bergson and William James, and John Dewey. For a while, he became a Fabian socialist, but he would not accept anyone's system of ideas for long. For him the only constant was change, not aimless flux but change to be mastered by the application of intelligence and made to rebuild an inefficient, often unjust society, driven by irrational forces, into a more humane and rational system.

Lippmann became an editor of a fledgling journal, the *New Republic*, at a time when its objects were to subject capitalism in the United States to enlightened centralized control, to promote American cultural nationalism, and to be radical but not socialistic, pragmatic rather than doctrinaire. In the *New Republic*'s comfortable offices, paid for by heirs to Morgan and Rockefeller money, he associated with such other pragmatic intellectuals as Herbert Croly (the journal's founder), John Dewey, Walter Weyl, Felix Frankfurter, and with distinguished national figures from politics, finance, and the law, including Louis Brandeis, Oliver Wendell Holmes, and Ogden Mills. These men admired the bright young writer, and he cultivated their friendship and used his connections with them as he sought to influence public affairs.

When the war broke out in Europe, Lippmann and his *New Republic* colleagues tried to chart a course for the United States. It was a shifting course. Neither he nor the other editors held any brief for the Central Powers. Lippmann, Croly, and Weyl believed a victory of Germany and Austria-Hungary would threaten the United States. Though they sympathized emotionally with England and France, they did not consider the Allies altogether blameless. They refused to accept the stories of German atrocities at face value and complained when the British interfered with American trade. Lippmann, who believed the war had originated in a struggle of imperial powers for control of less-developed areas, argued as late as 1916 that if the United States joined the Allies in a righteous cause, it would end by fighting to decide who controlled the Middle East. By the close of 1916, the *New Republic,* encouraged by Walter Lippmann, who felt the United States had a vital interest in having "the Atlantic Community" (mainly Great Britain and America) command the sea links between those countries, called for limited participation in the war to control Germany and establish a league of nations.

Before the election of 1916, President Wilson found it expedient to cultivate the editors of the *New Republic,* who, as former admirers of Theodore Roosevelt, had influence among liberals who had once supported Roosevelt and his Progressive party. The journal's editors, who earlier had regarded Woodrow Wilson's New Freedom as unrealistic for the era of modern capitalism, had recently found the president's domestic program and some of his appointments to their taste. It was arranged for Lippmann to visit the president. The two men exchanged ideas for nearly two hours, speaking about social reform and Wilson's principle of "benevolent neutrality." Following this interview, Lippmann persuaded the *New Republic* to support the president, who was running on the slogan "He Kept Us Out Of War," precisely because Wilson seemed willing to go to war.

Having brought the *New Republic* intelligentsia into his camp, Wilson attempted to control them. In December, Lippmann received an invitation to dinner at the White House. He joined the president's confidential advisor Colonel House at New York City's Pennsylvania Station and rode down to Washington with him in a private drawing-room car. Afterwards, House met with Lippmann and Croly once a week in the colonel's New York apartment to keep them "on the right road," as he told the president. In exchange, word of these meetings gave the journal and its editors the appearance of influence, an illusion fostered by the fact that Wilson used phrases like "peace without victory," which the *New Republic* had previously employed.

By the time the United States became a belligerent, Lippmann's qualms about the Allies had dissolved. Not only did he feel that America was in a "people's war," the object of which was a "union of liberal peoples"; he also contended that war would turn the United States into a social democracy. "We stand at the threshold of a collectivism," he wrote, "which is greater than any as yet planned by a socialist party." The war that had started as a clash of empires had turned (in Lippmann's mind) into a worldwide democratic revolution.

For some time, Walter Lippmann had ached to play a part in what he called "the supreme event of our lives." He was of military age and subject to the draft, but unlike his Ivy League compatriots who were rushing to enlist, he did not immediately seek a commission. Instead, he asked Felix Frankfurter, who worked in the War Department, to help him secure an exemption so he could spend his time "studying and speculating on the approaches to peace and the reaction from the peace." Frankfurter put in a word with Secretary Baker; then Lippmann wrote Baker that he had spoken with all the people whose advice he valued and that every one of them had urged him to apply for exemption. He was reluctant, he said, to ask a favor. "You can well understand that that is not a pleasant thing to do, and yet, after searching my soul as candidly as I know how, I am convinced that I can serve my bit much more effectively than as a private in the new armies." Lippmann explained to Baker that he would work at the job "with all my heart because I'd rather be under a man in whose whole view of life there is just the quality which alone can justify this high experience. I needn't tell you that I want nothing but the chance to serve, that salaries, titles, ambitions play no part whatsoever." He added that his father was dying and his mother was "absolutely alone in the world." (Lippmann's father survived for ten more years.)

The secretary of war welcomed Walter Lippmann to his staff and started him out by having him listen to his own ideas. Lippmann next went to the Cantonment Adjustment Commission as a specialist on army labor problems and then served as a War Department representative to a committee on wages in arsenals and navy yards. In the fall of 1917, Colonel House tapped him to join the Inquiry, a group of scholars who developed information the government used to prepare its positions on the peace settlement. Lippmann became its general secretary, coordinating the activities of 126 men and women, incorporating their data into reports for Colonel House and the president, and providing direction to the members who examined political questions. He helped draw up plans for postwar frontiers and contributed to what eventually became Wilson's statement of American war aims, the Fourteen Points.

Colonel House considered Lippmann invaluable in this job, a source of ideas, a conduit to the liberals the administration wished to cultivate, and one of the most influential and easy to get along with of the liberal group. House liked the fact that Lippmann kept his mouth shut when required. Discretion was a crucial quality for Inquiry staff since the administration had decided to keep the organization secret. Nevertheless, word of its activities got out, leaked, the president believed, by Jewish War Department aide Felix Frankfurter. Colonel House was no philo-Semite. He complained that "Jews from every tribe have descended in force, they seem determined to break in with a jimmy if they are not let in." But he doubted that Lippmann was the leaker: "The objection to Lippman[n] is that he is a Jew, but unlike other Jews," House told the president, "he is a silent one."

After working in the Inquiry for several months, Lippmann secured a commission from the War Department and traveled to Europe as a captain in charge of an army propaganda unit. At AEF headquarters, he wrote leaflets to be dropped behind enemy lines and traveled to Paris to have them printed. For several weeks, he interrogated captured Germans. Meanwhile, as Colonel House's representative to Allied intelligence offices, he tried to coordinate their work with the investigations of the Inquiry. An army colleague of Lippmann's described their life in those days: "We are quite blasé from meeting bigwigs. . . . We're going at it like those to the manor born. State secrets between glasses of Graves, that's the method."

Lippmann had made himself a splendid wartime career, serving the nation in several capacities, providing liaison for the government to other liberal intellectuals, and helping the administration prepare for the peace settlement. In Washington and in Europe, men at the center of affairs accepted him, after a fashion. Yet during these months, all was far from well with him.

Personal conflicts developed between Lippmann and another member of the Inquiry, the geographer Isaiah Bowman, who played an increasingly dominant role in the organization. Bowman complained to House that Lippmann was a bad influence on the Inquiry and that he tended to disorganize its work by taking people away from other tasks for his own purposes.

The administration's methods of managing public opinion abroad and in America distressed Lippmann. Like John Dewey, he took a pragmatic approach to the problem of managing information. When Herbert Croly complained about the way the administration was cracking down on dissenters, Lippmann told him that Croly could not afford to "express his vexation every week because the war is a brutal and unreasonable thing" and that if the *New Republic* was to influence American diplomacy, it would have to sound "as if it really believed in a vigorous fighting policy. . . ." To Colonel House, he declared that he had "no doctrinaire belief in free speech. In the interest of the war it is necessary to sacrifice some of it." Yet he disdained George Creel's approach to selling the war and objected to official assaults against the liberal press, telling House that the government's method was breaking down liberal support for the war and tending to divide the nation's "articulate opinion" into "fanatical jingoism and fanatical pacifism." At House's suggestion, he prepared a memo for the president, in which he argued that a great government ought to be "contemptuously uninterested" in socialist attacks and should censor only military secrets and advice to evade or break the law.

Wilson appreciated Walter Lippmann's help but not his criticism. The president was happy when Lippmann helped him influence public opinion, particularly the opinion of liberals. "Lippmann is not only thoughtful, but just and suggestive," he wrote Secretary Baker after Lippmann helped draft one particularly delicate statement, a reply to a peace proposal from the Pope. When Lippmann praised him, the president responded warmly. For instance, after the journalist had written the secretary of war that one of Wilson's speeches had made "the whole world" seem better and lauded the president's leadership, Wilson told Baker, "Lippmann's letters all make one like him." But when Captain Lippmann disparaged the work the Committee on Public Information was doing in Europe, the president remarked: "I am very much puzzled as to who sent Lippmann over to inquire into matters of propaganda. I have found his judgment most unsound, and therefore entirely unserviceable in matters of that sort because he, in common with the men of the *New Republic,* has ideas about the war and its purposes which are highly unorthodox from my point of view."

The men of the *New Republic* discovered before long that the administration that used them to ensure the adherence of liberals to its policies could not or would not secure the kind of peace settlement that they and other liberals wanted. Lippmann and Croly eventually decided that the Treaty of Versailles was a catastrophic failure and warned that the peace it provided for could not last and that Americans would be "fools" if they allowed themselves to be "embroiled in a system of European alliances."

Lippmann attributed the disaster to Woodrow Wilson. Like Dewey, he argued that the president should have insisted that the Allies agree to his program before the fighting ended (though he was no clearer than Dewey on how Wilson could have done this while holding together the sea links that joined the "Atlantic community"). He claimed that by acquiescing in the silencing of critics, Wilson had undermined public support for a better peace. "It is a very dark moment," he wrote Secretary of War Baker, "and the prospect of war and revolution throughout Europe is appalling."

After a while, Lippmann came to recognize that others, including himself, had contributed to the tragedy of those years, that events had eluded their control, that the morale he and other publicists had done so much to develop exhausted the nation's public spirit, and that the people of the wartime generation had been unreasonable and violent. More than a decade after the Armistice, he confessed, "If I had it to do all over again I would take the other side; we supplied the Battalion of Death with too much ammunition." Decades later, during the Vietnam war, he became a gadfly to a president. But in World War I, young Walter Lippmann, like

Dewey and other pragmatic liberals, learned that it was easier to imagine that war could be controlled than to do it.

Randolph Bourne, Goliath Slayer

Randolph Bourne had been saying that all along. As one who habitually deflated received ideas, Bourne took a special interest in puncturing the notion of his fellow intellectuals that the Great War could be made to serve liberal ends.

There were certain resemblances between Lippmann and Bourne. Like Lippmann, Randolph Bourne attended an Ivy League college (in his case Columbia) and became a non-Marxian socialist and a devotee of pragmatism and the experimental approach. For a while, he worked for the *New Republic*. The differences, however, between the two men are much more significant.

If Lippmann's heritage made it somewhat difficult for him to enter the American establishment, Bourne, the scion of a middle-class Protestant family, was much more the outsider. Life had struck him a series of blows, starting at birth when a clumsy forceps delivery scarred and permanently distorted his face. When he was four, tuberculosis of the spine twisted his back and stunted his growth. In Randolph's twelfth year, his mother's family forced his alcoholic father, an unsuccessful salesman, to leave home. These misfortunes did not prevent the young man from reveling in love and art, in youth and friendship and conversation; but they certainly shaped his perspective, making him an advocate for the exploited and the different, a relentless assailant of the conformist calculating bourgeoisie. Bourne wrote, for instance, a cynical attack in the *Atlantic* on his home-town, Bloomfield, New Jersey, sneering at its elitism and the crass manipulations of the factory owners and their families who dominated its society.

With the help of Charles Beard, one of his professors at Columbia, Bourne found work at the *New Republic*, writing dozens of essays and reviews and becoming its expert on progressive education. But he was never at home in its club-like surroundings. It was not just his high-pitched, scratchy, nervous voice or his small hunchbacked body that he covered with a flowing, ankle-length cape that alienated him from people there; it was his personality and his attitude. No matter what he accomplished, he could not feel he had done well enough. He craved reassurance and praise, which made his relations with fellow workers awkward. He seemed anxious about his position, justifiably, it turned out. The *New Republic*'s other bright young men were serious and a little stuffy. Bourne was cheerfully flippant, at times sophomoric, and he would not hide his dislike of self-important people.

When he joined the *New Republic,* Croly had told him: "We have got to be thoroughly critical, but there must be also a positive impulse behind our criticism." Bourne, however, had an instinct for irony, and his tone sometimes clashed with the journal's sane progressivism. Croly had also explained that to influence the people it was trying to reach, the *New Republic* needed to include a certain amount of conscious patriotism in its critical standards. Bourne was patriotic enough, but he made it clear in Croly's magazine and elsewhere that he found things to admire in other nations besides his own and that he detested parts of America's culture: the destructive individualism he observed in its rich inhabitants, its "frowzy towns, its unkempt countryside, its waste of life and resources, its stodgy pools of poverty."

Bourne's views of foreign countries, particularly England and Germany, differed considerably from those of his colleagues. During a trip to Europe in 1913, he had visited England, where much of what he saw distressed him. The architecture seemed generally ugly, the food unappetizing, and the people insular and stiff. He found it hard to discern what Englishmen really felt. Aware of how it felt to be excluded and controlled, he disliked the class-consciousness and imperialism of the English upper classes with their conviction that the lower elements deserved their lot. Militant English suffragists excited Bourne, and he admired some of the radical socialists he met. He recognized that England enjoyed a free government and humane religion; yet he was appalled by the thought that only a tiny minority of its people lived in what he considered decent circumstances.

Bourne was also troubled by the atmosphere in Germany that year, particularly by the tendency of Germans to believe in their government without questioning. He abhorred German militarism and objected to Germany's invasion of Belgium. Nevertheless, he respected and was much influenced by German culture. He coined English versions of German compound words, in particular, the words of Friedrich Nietzsche, a German philosopher who, like Randolph Bourne, criticized bourgeois values. In a 1915 *New Republic* article, "American Use for German Ideals," he observed that the Germans had produced town planning, beautiful public buildings, social welfare, and Nietzsche. Too much had been made of Germany's militarism and industrialism and not enough of its spiritual energy. England was decaying while the "true Germany" was a "peace state."

Such views, and Bourne's feeling that the United States could not possibly fight for democracy if it joined Great Britain in the war since Britain itself was not really democratic, put him at odds with the *New Republic* editors. After publishing "American Use for German Ideals," they offered him few significant assignments, and he began to write for other journals (the *Dial* and especially the *Seven Arts*) that tolerated his radicalism.

The *Seven Arts* began publishing in 1917, financed by a wealthy woman whose analyst had suggested she might improve her self-esteem by supporting a radical journal. The editors, James Oppenheim and Waldo Frank, knew that Bourne was extremely controversial and that he had helped to form a Committee for Democratic Control to call for a national referendum on whether America should enter the war. They realized he would write about politics as well as art; but they were courageous or foolhardy enough to hire him anyway and to let him publish a series of antiwar essays despite the increasing likelihood that Bourne's writing would lead the government to close them down.

Bourne aimed his pen at several targets. He scorned the young men who went to Europe and wrote about the heroism of the troops, the suffering of friendly civilians, and the nobility of the Allies. Minor novelists and minor poets and minor publicists, he said, were coming back from driving ambulances in France and writing books to persuade Americans of the "real meaning" of the war. It seemed an irony to Bourne that many of these writers, with their generous desire to serve and the pity and horror they felt at the suffering they saw in France, were unmoved by exploitation at home and by the arid quality of American life. Having never felt responsible for labor wars or racial exclusion and oppression of the masses in America, insensitive to "the horrors of capitalistic peace," these patriotic writers had a large store of idle emotional capital to invest in Europe's ravaged villages and oppressed peoples.

Bourne could understand why such people saw the war as they did. They were part of the least democratic segment in America, connected to the richer and older classes of the Atlantic Seaboard or to upper-class elements in other parts of the country that identified with an eastern ruling group. Their roots lay in the original centers of preparedness and pro-war agitation. But what really distressed him were the liberal intellectuals, the "herd intellect," he called them, who had joined the country's worst reactionaries in supporting Wilson's war.

In his college years, Bourne had worshipped one of the leading pro-war intellectuals, John Dewey. He had called the philosopher "a prophet dressed in the clothes of a professor of logic," promoted Dewey's theories of education in the *New Republic,* and remarked at one point, "We're all instrumentalists here." But as Dewey published essays in support of the war, Bourne attacked him with the feeling of a student betrayed by his teacher, lumping him with pro-war socialists and conservative labor-union leaders who, Bourne said, by accepting the war were living out Dewey's instrumental philosophy.* Dewey, he argued, had placed too much emphasis on technique for achieving objectives, not enough on values; he had preached adjustment to social conditions when adjustment could be fatal to one's goals. America had gotten into the war, Bourne declared, because its government "practiced a philosophy of adjustment."

Toward members of the younger generation who followed Dewey's lead, the "earnest group of young liberals" like Lippmann, Bourne expressed contempt. There seemed, he said, to be "a peculiar congeniality between the war and these men. It is as if the war and they had been waiting for each other." And it was John Dewey who had helped make them the way they were:

> The young men in Belgium, the officers' training corps, the young men being sucked into the councils at Washington and into war-organization everywhere, have among them a definite element, upon whom Dewey, as veteran philosopher, might well bestow a papal blessing. They have absorbed the secret of scientific method as applied to political administration.

These followers of America's best-known philosopher of education represented to Bourne a new force in American life, the product of a shift in the colleges from classical studies to a training that emphasized political and economic values. Their education had given them no coherent system of "large ideas," no feeling for democratic goals. Having learned too literally the instrumental approach, these energetic, "immensely intelligent" young men had become preoccupied with the technique of conducting war. With no distinct philosophy of life, except "intelligent service," and no clear conception of what kind of society America should be, they had made themselves efficient instruments for the government.

Applying to the war intellectuals the insights of the "new psychology"† with its emphasis on unconscious motivation, Bourne declared that the reasons they gave for supporting the government were simply rationalizations. The war had shocked them, shattering their belief in the way the world was moving and their hopes for democratic nationalism and humane internationalism. It aroused in them an itch to share in the rest of the world's great experience. The salve for their emotional trauma was to accept the war without questioning it. Spared from the mental discomfort of living with suspended judgment, avoiding the dangerous course of challenging the administration, the pro-war intellectuals floated with the current, enjoying the "peacefulness of being at war."

---

*The theory that ideas are plans of action to control the environment, not reflections of reality.
†This refers chiefly to the ideas of Sigmund Freud.

These so-called "realists," with their "nice sense of purposive social control," thought they could tame the war and make it serve their purposes. They believed it was better to enter the war and master it than to drift in blindly. But if the war were too strong for liberals to prevent, Bourne wondered, how could it be weak enough for liberals to control? Bourne felt the *New Republic* intellectuals were blind to the fact that the war was run by statesmen whose only motto was "win, then grab what you can." They appeared not to see that a Germany forced to accept democracy by the victorious Allies would remain a constant menace to the peace of Europe, that the League of Nations the war intellectuals thought they were fighting for would freeze the map of the world and serve as an instrument for America's financial imperialists.

Bourne prescribed a course for intellectuals who wanted to preserve their moral health: Remain aloof from the war. Neither accept it spiritually nor obstruct it. Live as a "spiritual vagabond," detached from society and from middle-class respectability. Focus on the personal, social, and artistic ideals that would enrich American civilization but find no consolation in the hope that something good would emerge from the war. Sneer at those who "buy the cheap emotion of sacrifice." Call for peace unceasingly, insist that the settlement really be democratic and contend against a new world order that, though supposedly liberal, was founded on military coalitions and allowed countries to retain the armaments for fighting another war.

He found it difficult to live according to his own advice. His heretical writings, his attacks on his intellectual mentor, his searing critique of onetime peers, his challenge to a government that pounced upon dissenters, especially less-respectable ones, embroiled him in controversy and intensified his feelings of isolation and difference. Always craving companionship, Bourne felt extremely lonely. "I seem to disagree on the war with every rational and benevolent person I meet," he declared, "I feel . . . very much out of touch with my times."

While the editors of the *Seven Arts* continued to stand behind him, his articles and contributions by the radical journalist John Reed brought warnings and severe criticism from other members of the American intelligentsia, including Van Wyck Brooks, Amy Lowell, and Robert Frost. Frost wrote:

> In the Dawn of Creation that morning
> I Remember I gave you fair warning:
> The Arts are but Six!
> You add Politics
> And the Seven will all die a-Bourneing.

The magazine's benefactor became increasingly distressed. People close to her warned her that she was assisting subversive pro-Germans. Friends shunned her. Her family talked about placing her in a mental institution. She ended her contributions and a few weeks later killed herself. A year after its first issue, the *Seven Arts* expired. There was still the *Dial*, but its editors wanted John Dewey to join them, and when Dewey told them he would do so only if they fired or demoted Bourne, they reduced him to book reviewer.

Still, he continued to write, musing in a piece called "The State" on the benefits war provided to the state and its citizens. War vivified the state, he said; it was "the health of the state," filling every cell of the body politic with activity. To the citizens, the "herd," it brought the most primitive kinds of psychological gratification, giving them a feeling of purpose, providing those who thought and felt like everyone else "the warm feeling of obedience, the soothing irresponsibility of protection," and a sense of power, while those outside the crowd felt helpless and forlorn. The state's ruling organization presented itself to the masses as a symbolic family—as kindly, protecting father, Uncle Sam, and as tender mother in the Red Cross posters. Before those images, the people at war became children again, obedient, respectful, trusting, full of naive faith in the all-powerful adults, who cared for them, imposed mild, but necessary discipline, and took away their anxieties.

Bourne died before he could finish "The State," a thirty-two year old casualty of the 1918 influenza epidemic. Another rebel, John Dos Passos, inscribed his epitaph:

> This little sparrowlike man, tiny twisted bit of flesh in a black cape, always in pain and ailing, put a pebble in his sling and hit Goliath square in the forehead with it.

> *War*, he wrote, *is the health of the state.*

What led these men to their differing views about the possibilities for controlling war? Logic, observation, their relationships to the culture of their time, and personal circumstances all played a part.

In the case of Bourne, the personal element seems especially strong. For all his friendships and his joy in life, Bourne still regarded himself as something of an alien. This attitude drew him into the camp of the enemies of respectability and authority and immunized him against the disease of pro-war liberal intellectuals—wishful thinking about the war. And while his alienation may have distorted his perception of certain things, it also gave him an insight into the wartime system sharper than those of people who took part in its operations—John Dewey, for example.

Hindsight allows us to ask how Dewey could have thought it possible to make the Great War yield the results he desired. Dewey was by no means naive about the way people and societies functioned in crises. Influenced, like Lippmann and Bourne, by the "new psychology," he had long recognized the power of emotions to overwhelm reason and understood how wartime emotionalism could warp rationality. In a 1916 essay, "On Understanding the Mind of Germany," Dewey observed that successful warfare depended on illusions and that when the fate of a nation was about to be decided, loyal people did not analyze very carefully the information reaching them. To make the horrors of war bearable, everyone had to feel that what his country did in war was morally justified, and justification for suffering had to be found in the ideals for which the war was fought. "Emotional perturbations," he wrote, were "so deep and general in war that anyone who keeps himself outside can behold the suborning [sic] of intelligence in process."

It seems fair to ask why, if Dewey thought war could do this kind of damage in Germany, he did not feel it would subvert intelligence in the United States. It also seems fair to wonder how he expected the Wilson administration to force the Allies to abandon their war aims and why he imagined the American people would suddenly abandon the idealism, the sentimentality, and the pious optimism to which he ascribed the war's unhappy outcome.

To reconstruct what made him support the war and led him to think it could be directed along lines he laid out, one has to imagine the forces that acted on him. He belonged to a class of Americans, members of Ivy League faculty, who from the outbreak of the war had been overwhelmingly pro-Ally, despite, or perhaps on account of, the fact that many of them had been educated in Germany. When the United States entered the conflict, pressure on university intellectuals to support the war was virtually irresistible—from fanatically pro-Allied students, from administrators and trustees, and from colleagues. There was also the special pressure that reformers felt. As the pacifist Jane Addams said, "We were constantly told by our friends that to stand aside from the war mood of the country was to surrender all possibility of future influence, that we were committing intellectual suicide. . . ."

Yet Dewey probably required little prodding. Like millions of other Americans, his patriotic heritage impelled him to support his country in the First World War, whatever his qualms. He had been born in the state that sent Ethan Allen and the Green Mountain Boys to fight in the Revolution. When he was an infant, his father left his family and business to go to Virginia with the Vermont Cavalry to put down the Rebellion. Afterwards, John Dewey heard his father reminisce about that terrible war which really had produced social change, a kind of change people like Dewey approved of. For John Dewey, war was a regrettable but normal part of life.

Dewey had believed before 1917 that imperial Germany menaced his country. A critic of German philosophy in prewar years, he blamed Germany's ruthlessness and warlike character on that country's predominant philosophical absolutism and warned the United States not to aid through inaction an empire that considered anything that helped it win permissible or even sacred. Just before the United States entered the war, he feared that unless it intervened, Germany would probably triumph, becoming a dangerous neighbor to America. Dewey thought that force could be used, intelligently, to curb "German lust for spiritual and political monopoly." He felt Wilson was justified in trying to prevent the Central Powers from winning. Feeling that the United States had to join in, Dewey concluded that it made sense to try to control the war's outcome.

Lippmann shared Dewey's geopolitical outlook and was subject to the pressures liberals experienced in the war years, but the personal factors that led him to try to shape the war differed from those that affected the philosopher. While Bourne was something of an alien and Dewey was squarely within an old WASP tradition, Lippmann was just beyond the margin, constantly attempting to become an insider. His path to influence during the First World War was to serve the administration as a house intellectual. To play this role, Lippmann had to suspend his cynicism about Allied war aims and to argue that the most destructive war in history could be directed toward the outcome he desired.

This is not to deny that, for a while at least, Lippmann, like Dewey, really thought the Great War could lead to a liberal reconstruction of the world. Their perspective was as different from ours as it was from Bourne's. They had just lived through an era of cultural transformation— in educational methods, in psychology, physics, philosophy, and the arts— a time when old forms were being broken and when it was possible to imagine that society could be remade, with social change subject to rational control, and that a liberal democratic world could emerge from the catastrophe of World War I. However fantastic this last point seemed years later, as it seemed then to Randolph Bourne, people did believe it in both Europe and the United States.

There are certain optimistic times when intelligent people can be readily deluded, even those who understand how emotions can deform rational thought. Such was the heady epoch near the beginning of the twentieth century when certain leading American intellectuals, influenced by personal circumstances, rationalized away their doubts, fit the Great War into the framework of their philosophy, and imagined that out of the violence of those years a better world would emerge.

9 ||||| The University at War: Veblen, Yerkes, Beard, and Cattell

The outbreak of war in 1914 created a painful conflict for American academics. Many of them had once looked to Germany as a center of advanced learning and culture. Several who had taken doctorates there fondly recalled their days in German university towns. But like most American university students and administrators, they generally sided with Great Britain and France, regarded the Central Powers as chiefly responsible for the war, and were appalled by the way Germany conducted it. Brander Matthews, a professor of dramatic literature, expressed the feelings of many American professors when he accused Germany of undermining its reputation for cultural eminence and noted that the Germans had destroyed art works in the university town of Louvain, bombed civilian centers, and retained a "barbaric medieval alphabet." When a group of German intellectuals (including the physicist Max Planck, the psychologist Wilhelm Wundt, and the biologist Ernst Haeckl) issued a defense of German actions, denying that German armies had done such things as intentionally pillaging Louvain, they angered Americans like Johns Hopkins philosopher Arthur O. Lovejoy, who wrote that the German professional class had failed to perform its proper function—"the function of detached criticism, of cool consideration, of insisting that facts, and all the relevant facts be known and faced."

When the United States joined in the conflict, college and university professors volunteered to help in whatever ways they could, even if this meant leaving their institutions. In fact, some longed to leave. James H. Breasted, Egyptologist at the University of Chicago, wrote: "I regret to say that we members of the Oriental Department have not yet been able to find enough to do in aiding the War to carry us away from the University."

These men expressed an attitude common in prewar American academic circles: that the university should serve society. But some of them were also afflicted by a kind of masculine mystique, an indefinable dissatisfaction with a life of scholarship and teaching while an exciting war was raging in Europe. The economist Richard T. Ely complained that he had missed the Spanish-American War, was too old to fight in this one, and felt "unsettled" at not being able to go. It was "painful to my wife as well as to myself," he wrote, "that I have not had a more active part . . . in this greatest war in the world's history." Archibald Henderson, a professor of literature at the University of North Carolina, kept "thinking of the trenches & death & duty & carnage. I am restless & feel dissatisfied. I wish I could get away to France." Ralph Barton Perry, a Harvard philosopher, announced to a group of academics, "[I]t is inglorious not to be *there*, as near the front, as *directly* taking part, as possible."

This is not to say that all professors wholeheartedly favored war with Germany. Many did not, and more than twenty were fired or forced to resign because they failed to support the war or were thought to lack enthusiasm for it. Some who doubted eventually became converts. Professor Breasted's son said of his father, a famed archeologist who had been educated in Germany and had formed professional attachments there:

> He was shocked to find his loyalty to these old German academic associations and friendships condemned as pro-Germanism by colleagues whose dispassionate, peace-time detachment had given way to hysterical bitterness and hatred; and was humiliated to find that his own powers of cool reasoning were themselves no longer immune to such corrosive influences. . . . Never before in his darkest moments had he suffered such anguish of mind.

Pressures for conformity, yearning for the battleground, a desire for excitement, patriotism, and an instinct to serve only partly explain the motivation of university people. They were rewarded, both emotionally and in material ways, for serving. For example, to provide college-trained officers for the AEF the government created a Student Army Training Corps at more than five hundred institutions. When enlistments and the draft threatened to empty the campuses, the SATC, subsidized with federal money, kept colleges and universities open and provided faculty with draft deferments and psychic rewards. Edward S. Corwin of Princeton observed; "The War Dept.'s need has saved our bacon as an institution, and for us individually our self respect, by giving us some useful work to do while drawing our salaries."

Part of that work was teaching a new curriculum, including a war-issues course designed (in the words of the course's national director) "to enhance the morale of the members of the corps by giving them an understanding of what the war is about and of the supreme importance to civilization of the cause for which we are fighting." A version used at Columbia evolved into the Western Civilization course later taught in hundreds of American colleges and universities.

Thus, the SATC provided the government with facilities for training officers. It kept the universities open during the emergency and enabled faculty to reform the university curriculum while making themselves useful to the nation. It exemplified the mutually profitable relationships that American professors formed with the United States government during the First World War.

Veblen: The Alien as Patriot

War service gave scholars and scientists from American universities a chance to test their ideas outside the ivory tower. Thorstein Veblen tried to seize that chance when he offered his talents to the United States government as a humble worker in its bureaucracy.

Although Veblen was born in rural Wisconsin and grew up in America, he acted and thought like someone from another universe. Nominally a professor and an economist, he went out of his way to profess to as few students as possible. His methods were largely those of anthropology. He spent his time dissolving conventions and abstract theories that disguised the way the world worked. To his readers and listeners, he tried to show why people behaved as they did in pursuit of wealth, power, and prestige.

Masking his subversive ideas with a façade of ironic, objective-sounding language, Veblen depicted America's wealthiest inhabitants as throwbacks to an earlier stage of evolution—predators, who through force and trickery had acquired great riches and who now demonstrated their success by displays of conspicuous waste, for instance, by dressing their women in elaborate clothes that rendered them incapable of work. Businessmen, whom ordinary Americans saw as heroes, Veblen regarded as saboteurs, less interested in creating new products and expanding production than in causing production to collapse so they could profit from shortages, keep wages down by sustaining unemployment, and buy up failed but potentially valuable enterprises. He claimed that middle- and working-class Americans, instead of seeking the greatest utility for their dollars or rebelling against those who enriched themselves at the expense of their employees, tried to copy the style of living of their exploiters.

Veblen himself lived as differently as possible from the leisure class. He seemed to emulate the poor Norwegian farmers from whom he had descended. His thick mustache and scraggly beard were as unkempt as his rumpled clothes, and he attached his socks to his pants with safety pins. At one time he lived in a basement that he entered and exited by crawling through a window. His furniture was built from boxes and burlap bags, and he declined to engage in such rituals as making his bed or washing his dishes, though after enough dishes accumulated he hosed them off in a tub. Veblen did not bathe excessively. When people came to visit, he would sometimes sit for hours saying nothing.

Although he had studied at the renowned graduate school of Johns Hopkins University and had secured a Ph.D. from Yale, had excellent recommendations from his teachers, and had useful connections (his wife, the daughter of a wealthy midwestern capitalist was the niece of the president of Carleton College, Veblen's alma mater), he could not find a position in the academic world. For years he just stayed at home loafing and tinkering, reading and thinking. In 1892, when he was thirty-five, the University of Chicago, which had just opened with a rich endowment from John D. Rockefeller and a group of wealthy Chicagoans (including the owners of the Marshall Field department store and the Swift meatpacking plants) gave him his first permanent job. He began to turn out a series of works, including *The Theory of the Leisure Class*, that established him as one of the world's least-orthodox economists. His manner of living, especially his liaisons with faculty wives and other women, led to his departure from Chicago after fourteen years, but he continued to produce remarkable books as well as articles and reviews that appeared in professional and popular journals.

By 1917 Veblen had become one of America's best-known intellectuals. H. L. Mencken wrote in the magazine *Smart Set* that "he was all over the *Nation,* the *Dial,* the *New Republic* and the rest of them, and his books and pamphlets began to pour from the presses, and the newspapers reported his every wink and whisper, and everybody who was anybody began gabbling about him. . . . Veblenism was shining in full brilliance. There were Veblenists, Veblen clubs, Veblen remedies for all the sorrows of the world. There were even, in Chicago, Veblen girls. . . ." Yet like other people whose public views departed from the sentiments expected of patriotic Americans, Veblen seemed headed for a clash with the United States government.

Veblen never tried to hide his disdain for war, for the groups that valued and profited from it and for the feelings that sustained it. Enthusiasm for war, he wrote in *The Theory of the Leisure Class,* was the index of a "predatory temper." It conflicted with productivity, concern for the welfare of the community, and efficient production of useful goods. By forcing obedience to the dominant cultural and political systems, war subverted the pursuit of truth.

In *An Inquiry into the Nature of Peace and the Terms of Its Perpetuation,* published just before the United States entered the conflict, he described patriotism as a survival from archaic times that enabled owners of property to prey on ordinary people by creating an emotional bond between the exploited and their exploiters. The buildup of arms and the wars to which arms competition led benefitted investing interests more than the ordinary citizen; yet under the guidance of piratical business classes, the common man, bewitched by thoughts of honor and patriotic feelings, supported arms and war and imagined that they profited him. "Patriotic sentiment," he wrote, "never has been known to rise . . . to the consummate pitch of enthusiastic abandon except when bent on some work of concentrated malevolence." It found its "full expression in no other outlet than wartime enterprise; its highest and final appeal is for the death, damage, discomfort and destruction of the party of the second part."

If these views were not enough to get Veblen into trouble, he also criticized British capitalism as wasteful and inefficient and refused, at the peak of the war hysteria, to contribute to the YMCA on the ground that it was an agency to defend the existing order. Nevertheless, to the amazement of friends and critics, Veblen firmly supported the war against Germany and volunteered to work for the Wilson administration.

Although the government needed a great many professionals to staff its war bureaucracies, federal agencies could not find a place for this illustrious scholar. In an attempt to secure a position on the Inquiry, Veblen submitted a series of memoranda about economic power and postwar international relations. The Inquiry was not interested enough to hire him. Walter Lippmann, its secretary, preferred that he write about how to stop countries that held trade concessions in less-developed areas from excluding other nations, such as the United States. But Veblen persisted, and in February 1918 he found a place in the statistical division of the Food Administration.

In light of his comments on war and patriotism, how do we explain Veblen's efforts to serve the war government? Had this extraordinary man, so alienated from the myths of American society, suddenly become enamored of his native land? Did he abandon his principles? To answer these questions, we have to look at how he regarded the country's chief enemy, at the proposals he submitted to the government while serving it or trying to serve it, and the whole context of his life and career.

In 1915 Veblen published *Imperial Germany and the Industrial Revolution*, an analysis of the rise of German power. He argued in this work that Germany had brought down from its past a set of institutions and a "popular habit of mind suitable to a coercive, centralized, and irresponsible control and to the pursuit of dynastic dominion." To these it had added the advanced technology of countries like England. The result was an efficient, ruthless state bent on dominating its neighbors. Much of Ger-

many's vaunted *Kultur,* Veblen said, was a "retarded" throwback to medieval and "submedieval" habits of thought. The German Empire, he felt,
would have to be defeated to ensure lasting peace.

Yet Veblen also criticized the Allies. Both sides, he said, had begun the
war deliberately, and British subjects were only "moderately" freer than
their German counterparts. Because it both upheld and deviated from the
official American line, *Imperial Germany* was simultaneously promoted by
the Committee on Public Information and banned from the mails by the
Post Office.

Veblen thought that neither an Allied victory nor the proposed league
of nations would prevent another war. All the belligerents were dominated
and would continue to be dominated by wealthy capitalists who would
disrupt the world economy and divide the nations to secure their own
objectives. Disarmament was impossible because the vested interests
would insist on maintaining armed forces and would continue to use
nationalism and patriotism to preserve their power at home and abroad. A
Wilsonian league would be nothing more than a combination of capitalist,
nationalist countries incapable of removing the causes of war.

Still, he saw a dim hope that if the war lasted long enough, it might
undermine those vested interests. Wartime privation would make ordinary
people more suspicious of wealthy elites and reduce the tendency to
emulate the rich that helped rich people dominate society. In wartime,
governments could not allow vested interests to conduct their usual kinds
of sabotage and had therefore shifted control of key industries to engineers
and workers who knew how to maximize production. While Veblen
thought this shift was only temporary and believed that the vested interests
had surrendered control only because they thought the war would bring
them long-term gains, he felt that now that the public had seen what
efficient management could do, they would resist giving industries back to
private owners.

As a bureaucrat in the Food Administration, Veblen offered suggestions to his superiors for improving American efficiency, for instance, a
stiff tax on employers of domestic servants so that able-bodied footmen
and butlers could be turned into stevedores and freight handlers. To avert
a labor shortage that threatened the nation's wheat supply, he recommended that the United States government stop prosecuting the IWW,
cease aiding and encouraging state and local authorities who harassed and
intimidated the Wobblies, and join forces with the IWW to create a mobile
army of workers that could be moved around to harvest grain. He suggested that the government to all intents and purposes abolish the commercial interests of country towns by nationalizing the food processing
industry, by taking over the mail-order business and strictly regulating

manufacturers who supplied the mail-order houses, and by absorbing retail and banking concerns into local post offices. These changes, he estimated, would eliminate 70 percent of the jobs performed in country towns, freeing bankers, merchants, middlemen, and others to save the crops. He informed his superiors, "it is fairly to be expected that the superfluous townsmen will put up with it all in a cheerful spirit of patriotic sacrifice for the common good."

His superiors filed these proposals. In the summer of 1918, convinced that vested interests dominated the Food Administration, Veblen resigned.

While it would be a mistake to reduce Thorstein Veblen's complex motives to a simple principle, he was clearly trying to use the war while contributing to it. While he aimed some blows at small-town commercial interests, which he had long despised, and wanted to help defeat the German Empire, he seems to have been playing a joke on the government. Veblen habitually bearded the lion in its lair. He had launched his arrows at the leisure class, ridiculed conspicuous consumption, and depicted the capitalist as saboteur while employed at a university funded by the Swifts and Rockefellers and by retailers of consumer goods. He had subverted the academic community he inhabited by writing a savage essay on the higher learning in America and the "captains of erudition" who administered it and by sleeping with his colleagues' wives. During the war, he served a nation he felt was dominated by vested interests by working in an agency headed by the great prophet of capitalism, Herbert Hoover, and offered proposals that would have outraged conservatives if they had understood what he was doing. Cloaked in patriotic service, camouflaging his aims with ironic language, Veblen bored from within.

Robert M. Yerkes and the Rise of Psychology in Wartime America

In a 1918 issue of *Science*, Professor J. S. Ames of the Johns Hopkins University physics department remarked about the war:

> For the first time in the history of science men who are devoting their lives to it have an immediate opportunity of proving their worth to their country. It is a wonderful moment; and the universities of this country are seizing it. The stimulus to scientific work is simply enormous; and the growth of our knowledge is astounding.

Much of what Ames said was clearly true. Scientists from American universities and from private institutions submerged themselves in war work, developing range-finding equipment and chemical-warfare devices, perfecting submarine detection equipment, and searching for ways to promote American and Allied victory through technology and science.

It might strike a contemporary reader as odd that scientists should have needed to prove they were useful. Decades of publicity have made their value both in peace and war obvious to Americans. But that was not so in 1917, at least not for certain fields like psychology. It was largely through war activities, and especially through publicity about those activities, that psychology (led by Robert M. Yerkes of Harvard University) found its place among the useful sciences in America.

A farm boy from Pennsylvania, Yerkes attended a local college and then went to Harvard, which awarded him a doctorate in 1902. He taught there in the philosophy department for several years, but his own interests were in evolutionary psychology, specifically in the intelligence and behavior of animals as they adapted to their environments. His ultimate goal was to create a science of man that could be used to predict and regulate human behavior and to reorganize human nature in a better way.

In pursuit of this objective, he developed tests of mental activity. To be sure that they worked, he had to try them out on a large group of subjects, a time-consuming and expensive task, particularly since he believed the tests should be administered to individuals, not *en masse*. He also wished to study the mental characteristics of primates, but his department declined to support that kind of work or to promote him, and he remained an assistant professor for years. He tried to induce a Rockefeller fund to underwrite his research—without success. However, a friend, Ernest Southard, a psychiatrist at a nearby state hospital, asked Yerkes to help him determine the intellectual capacities of his patients, and Yerkes began to work half-time with Dr. Southard, developing mental tests while continuing to teach in Cambridge. He was able to perform a few promising studies of the mental life of apes but could not continue for lack of funds. The year the United States entered the war, he was elected president of the American Psychological Association (APA). Yet his own career as a scientist seemed blocked.

It was a difficult time not only for Yerkes but also for psychology, which had not yet achieved wide acceptance in the United States either as a theoretical science or as a useful contributor to society. Psychologists had begun to earn Ph.D. degrees in Germany and the United States. In the 1880s they had formed their professional organization, the APA, and tried to set standards for membership in their discipline. They held meetings and published journals; they performed some research with practical applications in business and medicine, and they had made significant advances in testing techniques. But they were divided over what their subject matter ought to be and over the methods that could best advance understanding of the mind. According to one former APA president, they did not attract the ablest investigators; they had no endowed laboratory; and most of them taught at universities for poor salaries and did not have enough time and money for research. Psychologists generally lacked the sense that they belonged to a true profession with a status comparable, say, to physicians.

Yerkes recognized how the war could advance his discipline. In February 1916 he wrote the War Department, suggesting that it use psychological testing in its recruiting program. The day war was declared, he turned a meeting of psychologists into a rally that ended with him in charge of a committee to determine how psychology could contribute to mobilization. He also arranged to chair the Psychology Committee of the National Research Council, an agency that scientists had formed to coordinate their military activities. "The prospects are now excellent," he wrote to another psychologist, "that we shall have opportunity to do something important, unless perchance the war should suddenly end."

Yerkes sent Surgeon General William C. Gorgas a proposal to give tests to recruits that could help the army eliminate the mentally unfit and identify the intellectually superior. Gorgas liked the idea, and the army agreed to a trial program, but before this program could be carried out, Yerkes and his colleagues had to surmount certain obstacles.

The first problem was the psychologists' unreadiness to do what Yerkes proposed. They did not know how to measure emotional stability and, even after years of work on intelligence testing, had no instruments to determine the mental capacity of large groups of subjects in a precise, consistent manner. Since they could not establish the reliability of any tests they devised without trying them on a large number of recruits, the army would not know for weeks how useful the testing program was or whether it worked at all. Indeed, if the testers followed Yerkes's original ideas—to examine about a fifth of the recruits or possibly every one of them for about ten minutes apiece—the war might be over before the army received the results.

A second difficulty arose from the opposition of psychiatrists who were staking out their own role in the war effort and doubted that psychologists were competent to perform mental examinations. Unprepared to wage a territorial battle with the medical doctors, Yerkes grudgingly agreed to reduce his goal to classifying recruits through mass testing rather than individual interviews, and he offered to make the psychological examiners assistants to the psychiatrists.

The psychologists themselves, meanwhile, were split over his program. As soon as Yerkes laid it out, Walter Dill Scott and Walter Bingham, applied psychologists at the Carnegie Institute of Technology, became convinced that he was less interested in helping the army win the war than in using the emergency to advance psychological testing. Scott, who considered Yerkes's early proposal for individual testing absurd, became a consultant to the adjutant general's office, where he set up his own system for rapidly classifying army personnel through common-sense judgments.

Finally, although the surgeon general and other army medical officers were willing to let Yerkes proceed, the War Department almost destroyed his program accidentally when it was barely underway. The department directed commanding officers to submit evaluations of the work of psychological examiners. About ninety replied, almost all of them negatively. Some replies came from camps where psychological examiners had not yet arrived. The commanding officers apparently believed they were being asked to comment on psychiatrists, who had been evaluating their men for some time and who had recommended discharges for certain draftees on grounds of nervous or mental disorders. The commanders considered those recruits malingerers and blamed the psychologists for recommendations that psychiatrists had made. Yerkes eventually learned of this confusion, and he gradually persuaded many of the base commanders, medical officers, and even army psychiatrists that the psychologists were doing useful work.

The first phase of his testing program began in the summer of 1917. Yerkes was commissioned a major and affiliated with the surgeon general's office. Forty-one of his colleagues received civil-service appointments or commissions in the Sanitary Corps. His group devised preliminary tests, tried them on a few thousand men, and had the results analyzed at Columbia University. Then they adjusted the testing instruments (producing what they called the "Alpha test" for soldiers who could read and the "Beta test" for those considered illiterate) and distributed them to about three quarters of the army camps in the United States. Testers trained in a school of military psychology administered them to hundreds of thousands of men each month.

The Alpha test had eight parts and included arithmetic problems, identification of synonyms and antonyms, word analogies, number sequences, and questions to test common sense. The Beta test was pictorial and included questions that required the examinee to complete unfinished pictures or to compare forms. The results were graded on a scale of A to E. Psychological examiners administered these intelligence tests to over 42,000 officers and more than 1.6 million non commissioned officers and enlisted men. They also tested mental patients, malingerers, YMCA secretaries, conscientious objectors, German prisoners of war, applicants for civil-service jobs, and camp followers. On the basis of the results, the examiners recommended that more than 7,800 men be discharged, that over 10,000 be assigned to labor battalions because of low intelligence, and that 9,487 be sent to "development" battalions.

Several camp commanders considered the tests useful and made a certain score a prerequisite for officer training or asked for draftees in the higher categories. Others remained skeptical. An officer at Fort Dix remarked that one of his draftees, who had a very low intelligence rating, was "a model of loyalty, reliability, cheerfulness, and the spirit of serene and general helpfulness," and asked, "What do we care about his 'intelligence'?" A depot commander remarked that slower men with little education often made better soldiers than those with more "flashy minds" who would probably be rated higher. Officers complained that college graduates were trying to get out of the army by simulating feebleminded ness on the tests, that illiterate men who made low scores on the Beta test became capable soldiers when taught to read, that the examiners were siphoning off the best troops from some units, or that intelligence tests were superfluous since army psychiatrists were already eliminating the mentally incompetent. The army War Plans Division feared that testing would actually reduce military efficiency by inciting soldiers to ridicule troops identified as feebleminded.

Some of the commanders suspected, along with dissenting psychologists like Walter Dill Scott, that the intelligence-testing project was chiefly aimed at acquiring scientific data, not at helping the army. When examiners asked for information that seemed unnecessary to classifying troops on the basis of mental capacity—for instance, for the soldier's hometown and his state or country of birth—they strengthened these suspicions. The commander of Camp Meade declared that while the work of the psychologists would be "of great scientific interest at some future time," the intelligence tests had very little if any value in raising and training an army of draftees. An investigator for the General Staff, Colonel R. J. Burt, who tended to favor the project, warned that it had to be firmly controlled so that "no theorist may . . . ride it as a hobby for the purpose of obtaining data for research work and the future benefit of the human race."

These officers may have overstated the scientific value of the tests. There were many errors in scoring them, reputedly 25 percent at Camp Devens, and the tests themselves contained fundamental flaws. Yerkes himself knew that sex, race, language, and social and economic factors affected performance on intelligence tests. He had preferred to examine people one by one in a way that would take social and biological variations into account. But when this proved unfeasible, he administered a program in which the examining instruments were not only shaped by the requirements of mass testing but were culturally biased. Thus the Alpha test asked literate soldiers whether Scrooge appeared in *Vanity Fair, A Christ-mas Carol, Romolo,* or *Henry IV;* whether the Knight engine powered the Packard, Stearns, Lozier, or Pierce Arrow motor car; whether the unit of electromotive force was the volt, the watt, the ampere, or the ohm; and whether a Percheron was a horse, cow, sheep, or goat. The Beta test showed illiterate troops pictures of a camel without a hump and a tennis court without a net and asked them what was missing.

Yerkes claimed that these tests measured native intelligence. But even if that were so, the psychologists did not show that the results predicted how soldiers would act on the battleground or how well they would do their jobs behind the lines. In other words, they did not demonstrate whether the intelligence-testing program contributed in any significant way to the army. The psychologists claimed that high scorers would make good officers, that men who scored *C* or *C+* were noncommissioned-officer material, and that lower scorers were suited to the lower ranks; but they never proved that this was so, and there is no evidence that the army followed their recommendations consistently.

Still, the wartime intelligence tests did yield important results, if not the ones Yerkes promised. To those who believed in their validity and in certain theories about racial differences, they appeared to prove that blacks, who tended to score lower on average than whites, had less intelligence than whites, although blacks from such northern states as Ohio and Illinois scored higher than whites from such states as Mississippi and Arkansas. They were interpreted to demonstrate that immigrants from southern and eastern Europe were inherently stupider than people who had come from countries in the north and west of that continent. They contributed to popular culture the notion that millions of Americans had a "mental age" of thirteen or less, barely above the level of "feeblemindedness," thus increasing alarm about mental deterioration and what was called in those days race suicide. And they had a substantial impact on the profession of psychology, on the teaching of psychology in American universities, on the way knowledge and mental capacity were measured, and on the career of Dr. Yerkes.

Publicity about the army testing program overshadowed its demonstrated achievements. Yerkes was an able publicist. As chairman of the National Research Council Information Service, he and his co-workers advertised their war work in books, articles, and speeches, distributing thousands of copies of the pamphlet "Army Mental Testing." In his "Report of Activities of Psychology Committee, National Research Council, 1918," Yerkes wrote that psychology had helped to win the war, "incidentally" established itself among the sciences, and "demonstrated its right to serious consideration in . . . human engineering." "Wartime publicity," he declared, "accomplished what decades of academic research and teaching could not have equaled."

This last claim was probably true. Although the program produced no significant new psychological principles and little of benefit to the army, its publicity during the war led to a fad for testing as schools and colleges, businesses, and government agencies began using short-answer tests to place people into categories. Americans began to consider psychology a useful profession. As the noted psychologist James McKeen Cattell observed, "The army intelligence tests have put psychology on the map of the United States, extending in some cases beyond these limits into fairyland."

Psychologists became so involved in wartime projects like the army testing program that experimental work declined in the universities. But in the long run, the war helped their profession because it stimulated interest in their work, helped them develop networks, and encouraged private foundations to view them as a genuine science worthy of financial assistance. War activities raised their esteem in the eyes of other scientists who now perceived psychology as a legitimate discipline with special knowledge and methodology. The psychologists' new status opened more jobs to them and led more people to want to become psychologists.

None gained more from the war than Professor Yerkes, who ventured out of the university, acquired a vast body of test data at government expense, and, by disseminating the fiction that the army testing program had been a practical success, boosted his prospects as researcher-entrepreneur. Yerkes continued to serve on influential National Research Council committees and secured a Rockefeller grant to construct a group intelligence test that was administered to millions of children during the 1920s. Rockefeller money also enabled him to run an Institute of Psychology at Yale, where he became the country's leading expert on primates. Thus he used the war to advance a once-stymied career while promoting his discipline and what he, other psychologists, and much of the American public regarded as the public welfare.

Charles A. Beard and the Historians

While university professors like Yerkes used the war to promote their disciplines and advance their careers, other scholars, including the eminent historian Charles A. Beard, tried to serve their country in a more purely altruistic manner. Always a committed advocate of the war, Beard eventually decided to leave his university because of the way it treated colleagues less devoted to the war than he was.

Like other professionals, academic historians placed the tools of their discipline at the nation's disposal. They worked with the Committee on Public Information and the National Security League and through their own National Board for Historical Service (NBHS), promulgating their government's views of the causes and objectives of the war. Using the forms of scholarship, they attempted to make government messages believable. At the same time, they tried to demonstrate their value to society. Their object, according to the managing editor of the *American Historical Review,* was to create a powerful, homogeneous public opinion enlightened by history. Historians created a war bibliography, compiled war poetry and prose, and offered guidance and information to college and secondary-school teachers. With their knowledge of languages, they supplied the government with intelligence gathered from the foreign press about developments in Germany and Austria-Hungary and about possible violations of the Espionage Act in ethnic newspapers. Two historians (one of whom knew no Russian) authenticated a set of mostly forged documents purporting to show that Bolshevik revolutionaries were simply tools of the German government. Another team employed historical research techniques in an attempt to prove that Senator Robert M. La Follette of Wisconsin, an opponent of the war, had given aid and comfort to the enemy.

While most historians offered their service to the government willingly, others took part only under psychological duress. Thus Charles Breasted, who anguished over the hysterical atmosphere in which he found himself, wrote propaganda for the National Security League. Some questioned whether they should be writing propaganda at all, even for a noble cause. Frederick Jackson Turner argued that historians had a trust to maintain "as the ministers of historic truth." Yet Turner was willing to employ historical data selectively, presenting facts about past sufferings of Americans to "hearten us for our own."

Cornell University historian Carl Becker hesitated to falsify American life. Responding to an appeal from the Committee on Public Information for a Fourth of July address to foreign-born workers, Becker warned against trying to contrast "the horrors of German autocracy and the shining virtues of American and Allied democracy." America had never been less democratic in its industrial and economic organization, he declared, and the foreign-born who worked for the great corporations understood this. If they were told about American ideals of freedom, they would only think of the "shameless exploitation" of labor, government corruption, and the "ridiculous" gap between wealth and poverty.

Other scholars had no qualms at all about writing propaganda in the guise of history. Professor Samuel B. Harding of Indiana University produced a syllabus for history teachers—"The Study of the Great War: A

Topical Outline, with Extensive Quotation and Reading References." Published as a CPI pamphlet, Harding's syllabus was essentially an attack on the Central Powers. It included more than five hundred references to more than a hundred sources. Over a third of these sources were other CPI pamphlets; another third were diplomatic histories of the war prepared by the belligerents, with the documents arranged to favor the arguments of the Allies. When Professor Harding received a memorandum reminding members of NBHS high-school curriculum committees that their objective should not be to inculcate particular views, he complained that this suggested history teachers should be neutral. "In the present emergency," he said, " . . . if the teacher cannot conscientiously and wholeheartedly lend his influence to supporting the war with Germany, . . . he ought at least to keep silent."

There was no need for Charles Beard to keep silent. His position on the war generally accorded with the government's and with the outlook of most of his colleagues. But this was not because he had adapted to the wartime climate, for he was too proud and independent to shape his views to prevailing opinions.

Beard's forebears were hardworking practical Quakers, some of them Radical Republicans. By the time he was born in 1874, his father had become rich as a farmer, contractor, banker, and land speculator in Knightstown, Indiana. Charles went to a Quaker academy for a while, but it expelled him for printing an attack on the faculty and administration of Indiana University. As a student at De Pauw University, he began to read the works of social thinkers, and during the depression of 1893, especially during visits to Chicago, he became aware of how the poorer side of the nation lived. Social injustice in the land of equality aroused the conscience of this young Republican from the heartland of America, and when he went overseas for more education, he noticed how much more solicitous Germany seemed to be for the social welfare of its people. In England, influenced by the ideas of John Ruskin (a British reformer of the previous generation), Beard helped found a school for workers, Ruskin Hall, which opened in Oxford in 1899. He served as the school's secretary and lectured to workers about the history and social effects of the industrial revolution.

Five years later, he was back in the United States, where, after completing a doctoral dissertation on an esoteric subject in British history, he joined the Columbia history department. In 1907 he was appointed to a chair in politics and government. Students admired this simply dressed, straightforward professor with striking China-blue eyes and a fine voice, who spoke with passionate conviction. The faculty of his department thought well of their prolific colleague and eventually made him chairman.

Beard, the scholar–social reformer, became an expert on municipal

government and an advisor to those who tried to improve the management
of New York City. Scholarship, especially history, was a way for him to
enlighten the present, to provide a realistic view of how people and institutions operated so that his contemporaries, freed from illusions, would have
a better chance to improve their own society. He repeatedly observed how
economic relationships shaped events, an obvious notion to the son of a
banker and land speculator. Some of his most famous works, such as *An
Economic Interpretation of the Constitution,* depicted the founders and
leaders of the United States as practical human beings, affected by such
ordinary concerns as how to safeguard their property. Viewed in this light,
the Constitution lost some of its mythic quality; but Beard's interpretation
helped people understand how, when conservative judges used the Constitution to uphold the rich and powerful, they were interpreting words of
men, not expressing the voice of God.

Beard's travels in Germany had left him with no love of Prussian
militarism. He sided with the British and French in the World War; and
having no moral objection to warfare, which he once called a "tremen-
dous" factor in "the progress of mankind," he announced after the
Lusitania was sunk that he favored American intervention. Millions of
Americans, he said in a 1916 speech, would give their lives to prevent a
Prussian state in their own country. When Germany resumed unrestricted
submarine attacks and others spoke of armed neutrality, Beard insisted
that the United States must align itself with the Allies and help "eliminate
Prussianism from the earth." After the Germans had sunk an American
ship, he told a class, with tears in his eyes, that it was no longer possible for
the United States to stay out of the war and that German autocracy would
have to be destroyed.

Beard threw himself into war work. He had a farm in Connecticut, and
he offered to coordinate war activities by the farmers of his state. Like
most of his colleagues at Columbia and elsewhere, he was anxious to do
something useful for his country and placed himself at the government's
disposal. He contributed entries to the Committee on Public Information's *War Cyclopedia,* including one on atrocities that mentioned such
incidents of terror by the Central Powers as German depredations in
Belgium and the Turkish killing of Armenians. The sources for this article
were official French and British investigations.

Like Carl Becker, Beard found the idealistic phrases of American
propaganda distasteful and possibly damaging to the cause. He wanted his
nation to approach the war realistically. In a June 1917 *New Republic*
article, he called for "cold-bloodedness, and a Machiavellian disposition to
see things as they are and to deal with them as they are—whether we like
them or not." A "painful consciousness" of the "purity of our purposes"

and of the rightness of American intentions, he said, would probably not be helpful. The United States and the Allies had to win over the Russian revolutionaries and the German Social Democrats, which meant persuading them that the war was not fundamentally or potentially a capitalist struggle for colonies, markets, and concessions. This would not be easy because the British had made it clear that they intended to keep German colonies and because the United States had its own imperialist record in Asia and the Caribbean. While slogans like "liberty against autocracy" might deceive Americans, they would not persuade the Germans because the Germans knew the historical record too well. Instead, the United States should inform the Germans that the American people would not "shed one drop of blood to enlarge the British empire."

Even after exposure of secret treaties confirmed his suspicions of the Allies, he continued to feel that the United States was fighting a just war. Yet the things that were done in the name of the war at the university where he taught troubled him greatly. To understand why, we must look into the events that led to the firing of one of his colleagues, James McKeen Cattell.

The Case of Professor Cattell

At the time of his removal from Columbia, Cattell was one of the most widely known scientists in America. Born to a prosperous Pennsylvania family in 1880, he was tutored at home and entered Lafayette College, where his father was a professor and later the president. During a visit to Europe, he spent a semester at the University of Göttingen and then went on to Leipzig, where he heard Wilhelm Wundt lecture on psychology. With a fellowship in philosophy from the Johns Hopkins University, he did research in that school's physiology lab. He began experimenting personally with opium, hashish, morphine, and other drugs for their effects and out of an interest in what they did to the mind. After a year, he lost his fellowship to another graduate student, John Dewey, partly because of his continual bickering with his professors and the university president and partly because Dewey seemed the better philosopher. He returned to Leipzig and became the first American to receive a doctorate in experimental psychology under Professor Wundt. His subject was a series of experiments on reaction times. Moving to St. John's College, Cambridge, he became intrigued by the work of Francis Galton, a scientist who studied physical and physiological variations among human beings.

Cattell pioneered psychological testing in the United States. He brought his research in human psychological measurement to the University of Pennsylvania, where he collected data on how different people performed on "mental tests" (a term he invented). For instance, he tried to find out how many letters a subject could repeat after one hearing and the time it took to respond to a sound. In 1891 Cattell was hired by Columbia University and established its psychological laboratory. Three years later, he began to gather physical and psychological measurements of all students entering Columbia College and the Columbia School of Mines—a project that went on for years. While he declined to offer an interpretation of the data he was collecting, he assumed that the facts would display certain patterns, for instance, relationships between the measurements and academic performance; but unfortunately for Cattell, his data refused to reveal what he wished. He knew little mathematics. When someone who did tried to determine if there were statistical relationships between the test results and the quality of student work, the correlations proved negligible.

By now Cattell was recognized as one of the leading experimental psychologists in the country. He had been president of the American Psychological Association in 1895 and six years later became the first psychologist elected to the National Academy of Sciences. But when his studies at Columbia failed, he abandoned experimental psychology and entered a new phase in his career that made him even more famous.

Professor Cattell had an independent source of wealth, which he used to buy and run scientific journals. Beginning in 1895, he edited *Science*, which became the journal of the American Association for the Advance of Science. From 1900 to 1915, he published *Popular Science Monthly* and (beginning in 1915) *Scientific Monthly*. He helped found *Psychological Review* and co-edited it from 1894 through 1905. He edited *American Naturalist* (from 1907) and (from 1915) *School and Society*.

Cattell took a special interest in the structure and social psychology of higher education, publishing a book on the subject in 1913 entitled *University Control*. He characterized the American university as an autocracy ruled by businessmen-trustees, who were ignorant of educational matters, and by presidents those trustees appointed. He wanted to democratize the institution by having faculty manage it, electing the presidents, the salaries of whom were not to exceed their own. But Cattell also criticized the faculty and felt they were partly responsible for their own debasement. The autocrats who ruled the university had reduced these "clerks" of the administration, as he called them, to a state of humiliating subservience and dependency. A symptom of the faculty's servile condition, and one of its causes, was the code of "gentlemanly conduct" under which professors operated. Cattell urged them to abandon this false collegiality and be truthful to one another and to the outside world. The university liked to keep its conflicts private. Cattell demanded complete publicity of its internal affairs, preferring, he said, to wash dirty linen in public than to wear it.

As an educational reformer, Cattell tried to live by his principles, saying what he felt and using irony and ridicule to attack the artificial gentility that surrounded him. His blunt behavior and his analysis of the university naturally angered faculty and administrators, some of whom thought he was deranged. Colleagues described him as "irretrievably nasty," "irrational," and "ungentlemanly." At one point, a university committee announced that it was impossible for Professor Cattell "to respect the ordinary decencies of intercourse among gentlemen."

At Columbia, as at Johns Hopkins, Cattell fought a running battle with the president—Nicholas Murray Butler—whom he referred to in a memorandum to the faculty as "many talented and much climbing." Butler and the trustees tried several times to have him silenced or removed. In 1913, after *University Control* appeared, the trustees were about to force him to retire, but a group of Cattell's colleagues prevailed on them not to do it. Four years later, after Butler announced that he intended to convert the Faculty Club building to other purposes, Cattell sent out a confidential memo to his colleagues proposing that Butler's house be confiscated for use by the faculty. The trustees again prepared to remove him but backed off when he apologized. When a group of professors became upset over the way he dealt with President Butler, he told them they felt that way because of the "traditionalism and the conventionalism, the lack of perspective and the lack of humor, which deaden university life."

The war finally brought him down. Cattell, who despised German militarism, had urged a negotiated peace; this put him even more at odds with the Columbia trustees, one of whom, Frederick R. Coudert, was legal counsel to the British embassy. After war was declared, Cattell went along with it, joining the Psychology Committee of the National Research Council. As an official of the American Association for the Advancement of Science, he helped mobilize the psychology profession. But this was not enough to satisfy the trustees or Nicholas Murray Butler—a onetime opponent of war who announced in a June 1917 commencement day address that "what had been tolerated before becomes intolerable now. What had been wrongheadedness was now sedition. What had been folly was now treason." There was no place in the university, Butler declared, for anyone who spoke or wrote treason. "This is the University's last and only warning to any among us . . . who are not with whole heart and mind and strength committed to fight with us to make the world safe for democracy."

During the summer of 1917, Cattell sent three congressmen a petition urging them not to approve a bill that would have allowed American draftees to fight in Europe. After one of the congressmen reported this to President Butler, the clerk of the trustees exclaimed: "We have got the rascal this time!" On October 1, the trustees fired Cattell for actions tending to foster disloyalty. The same day, they announced the removal of Henry Wadsworth Longfellow Dana, an antiwar English professor. Meanwhile, Butler blocked the rehiring of Leon Fraser, a pacifist. Earlier, Butler had forced the Political Science Department to employ Fraser, specifically to work for an antiwar group, the Association of International Conciliation.

At the time of these firings, an organization existed, the American Association of University Professors, that had been established to uphold the right of faculty to speak freely. John Dewey and Cattell himself had represented Columbia on a committee that founded the AAUP, and Dewey had served as its first president. (Beard refused to join it because he considered it a "futile enterprize.") In 1915 the AAUP's Committee on Academic Freedom and Academic Tenure had said that university faculty could only discharge their duties to the public if they were guaranteed "absolute freedom of thought, of inquiry, of discussion and of teaching"; it added that this pronouncement applied to what professors did outside the university's walls. One of the two main authors of this statement was Arthur O. Lovejoy, who had attacked German intellectuals for failing to offer detached criticism of their own government.

Two years later, the AAUP looked at things differently. A three-member Committee on Academic Freedom in Wartime, again including Lovejoy, now stated that college professors could legitimately be fired because of their "attitude or conduct" related to the war. For instance, they could be let go for knowingly uttering certain words whose "natural tendency" would be to stir resistance to the draft. Even if no legal proceedings had begun and guilt was unproven, such people could be dismissed. The AAUP committee also declared that German or Austrian professors who made hostile or offensive statements about the United States or its government, either in public or in private conversations with colleagues, neighbors, or students, should be removed from their posts.

If Cattell had to depend on the American Association of University Professors to get his job back, his cause was obviously hopeless. But his colleagues had saved him earlier. Could they rescue him in 1917?

Shortly before the declaration of war, Columbia's University Council had set up a joint faculty-administration committee, the Committee of Nine, to cooperate with a trustees' group that was about to investigate what Columbia's professors were teaching. One of the committee members was Dewey, whom Cattell had been instrumental in hiring twelve years earlier. Since then, the two men had fought repeatedly. The Committee of Nine took up the Cattell and Dana cases, recommending that Cattell, the advocate of complete publicity, be retired without telling outsiders the reasons and given a leave without pay for the rest of the year. Instead, the trustees simply dismissed him and issued an announcement to the press. They ignored the committee's view on Dana altogether.

Several of Columbia's professors were glad to see these men go, and eight of them, from the schools of Mines, Engineering, and Chemistry, went on record in support of the trustees. John Dewey took a different position. Because Dana had been forced out without a hearing by his peers, Dewey declared that no one could serve on the Committee of Nine and retain professional self-respect, and he resigned from it. Writing in the *New Republic,* he obliquely criticized Nicholas Murray Butler, remarking that "[W]e should have to go to German professordom" to find an analogy to what the trustees' supporters on the Columbia faculty had done.

One of his colleagues was more blunt. Charles Beard had had his own run-ins with the Columbia trustees. They had called him before them for supposedly condoning an insult to the American flag. They had accused him of teaching in a way "calculated to inculcate disrespect for American institutions," specifically for the United States Supreme Court, and had warned him to tell the instructors in his department that such teachings would not be tolerated. Later, when they announced their plan to investigate what was being taught, Beard led his department in refusing to cooperate. The behavior of Columbia's president and trustees in firing Dana and Cattell (and, perhaps even more, in firing Fraser for adhering to principles Butler himself had once espoused), outraged him.

A week after Cattell and Dana were ousted, Beard sent President Butler a letter. A "small and active group of trustees," he declared, who had "no standing in the world of education, who are reactionary and visionless in politics, narrow and mediaeval in religion" dominated the university. These "obscure and wilful trustees" terrorized the young instructors. Their actions could not be ignored because "we are in the midst of a great war and we stand on the threshold of an era which will call for all the emancipated thinking that America can command." He reminded Butler that he had always believed a German victory would "plunge all of us into the black night of military barbarism" and that he had been among the first to urge a declaration of war. He believed the war must be pressed to its conclusion. But thousands of his countrymen did not share his views and their opinions could not be changed "by curses and bludgeons. Arguments addressed to their reason and understanding are our best hope."

Beard did not want anyone to feel that he supported the war because he was in the pay of the Columbia trustees. As he put it, the arguments for the war had to come from men "whose disinterestedness is above all suspicion, whose independence is beyond all doubt, and whose devotion is to the whole country, as distinguished from any single class or group." He resigned from the university effective the following day.

The next day, he announced to his class that he had given his last lecture at Columbia. The students, silent at first, rose and applauded for over fifteen minutes while Beard wept.

As the United States government mobilized the country's scholars and scientists, people in the universities tried to use the war to secure their own objectives—the president and trustees of Columbia University, with the encouragement of certain faculty, to rid themselves of dissidents and gadflies; professors like Yerkes to advance their disciplines and their careers; and Veblen to enjoy the pleasure of offering iconoclastic ideas to government officials. The war was turned to advantage by professors who found in it a way to reform the curriculum or to feel part of a great event that would add excitement to their lives. At the same time, there were members of American university faculties who believed deeply in the objectives of the war and individuals like Charles Beard who acted without calculating the gains war service might bring themselves and their profession. Some professors dissented in silence. A few openly resisted and were punished. All were affected by wartime passions. In these ways, the university mirrored the nation at war.

10 ‖ The Battleground

In 1965 the historian William L. Langer marveled that after nearly three years of agonizing trench warfare, after Verdun and the Somme and other gruesome battles, all reported in excruciating detail, any American should have joined in the Great War without duress. Professor Langer wrote from firsthand knowledge. Forty-seven years earlier, he had been Sergeant Langer of the First Gas Regiment, American Expeditionary Force, serving on the battlefields of France. His government had given considerable thought at that time to the problem he referred to—how to induce Americans to enter and fight on the Western Front where millions of the Europeans had already suffered death and mutilation. The way the United States government managed that problem is the subject of the following chapters, beginning with a description of the battleground taken from accounts by American soldiers who fought there.

Before the United States entered the war, President Wilson concluded that he would have to send an army to Europe. At first he thought of raising a force of volunteers, with a draft to follow if more troops were needed, but he quickly realized that a voluntary army presented serious drawbacks. Since the scale of warfare required that the whole nation mobilize, it was vital to keep certain men of military age in civilian jobs essential to the war economy; yet these very men might volunteer to fight in Europe. A volunteer army would also have severe military disadvantages. Former president Theodore Roosevelt wished desperately to lead a division of volunteers to France. If the administration let him, the immensely popular onetime colonel of the Rough Riders, a leader of the opposition party, might take the army's best officers under his command when the rapidly growing AEF needed every available seasoned leader to train its recruits and strengthen its battalions. The political effects of a volunteer division led by Roosevelt were easy to envision—the publicity General Roosevelt would evoke, the trouble he could make for the commander in chief, and the votes he might win for the Republican party. In

Wilson's estimation, the wisest course was to abandon the notion of a volunteer army and start conscription at once.

By war's end, 72 percent of the four million men who served in the U.S. Army and over half the American Expeditionary Force were draftees; the rest were regular-army veterans, members of National Guard units, and individual volunteers. After a few months training (though some got none at all), two million of them sailed to Europe, where the commander of the AEF, General of the Armies John J. Pershing, began to turn them into an effective fighting force.

Pershing's ideas of what this army should do and when it should do it differed sharply from the views of the Allies. As soon as the United States was in the war, the British and French asked for a force of American regulars to bolster Allied spirits and demoralize the enemy. To satisfy their request, the American First Division, regular army, though filled with raw recruits, sailed in June 1917. The French and British wanted to use them and the Americans who followed to shore up their own weakened forces. With the backing of his commander in chief, Pershing refused. He insisted that the United States deploy its own army using its own methods of warfare. He reckoned that the American people wanted their sons to fight under the American flag, and he understood that an independent army would also strengthen the president's diplomatic hand.

The first units of Americans joined veteran British and French troops in quiet zones where the "Sammies" or "doughboys," as the Americans were called, could acquire a feeling for combat. But in March 1918, with Russia out of the war and Italy staggering from a defeat at Caporetto, the Germans launched a series of powerful offensives, opening a large gap in the British lines and driving the French back toward Paris. The situation became so threatening that Pershing had no choice but to send his soldiers where the Allies needed them. But by war's end, most United States divisions were fighting together.

As the Americans poured into Europe, they enlarged their combat role. During the German offensives, which lasted until mid-July, American forces helped blunt an enemy attack that crossed the Aisne River and reached the Marne in the vicinity of Château-Thierry, less than forty miles from Paris. It was there in early June that soldiers and marines of the Second Division counterattacked through Belleau Wood, taking and inflicting heavy casualties. Then, when the Allies went on the offensive through the area between the Aisne and Marne rivers, Pershing assembled a group of his best divisions and drove the Germans back, pushing them from a salient in the Allied lines at Saint-Mihiel, south of the Verdun battlefield. Beginning September 26, as part of an Allied offensive all along the Western Front, the American army, grown to six hundred thousand troops, battered its way through German fortifications from the Meuse River through the Argonne Forest. They continued to fight for forty-seven days until the Armistice silenced their guns.

Although American soldiers expended their lives freely, by European standards the cost in American blood was minuscule, less for the entire war than the toll of a single battle (such as the Somme or Verdun) between European armies. The armies the Americans faced had been drained by years of combat. Still, the moral impact of these fresh troops from the New World on friends and enemies and the prospect of millions more behind them multiplied the effects of American fighting power. Meanwhile, participation in the Great War made an ineradicable impression on the men who crossed the Atlantic to fight.

Nobody ever polled those men about what they were going through and how they felt about it, so it is impossible to talk of "typical" or representative experiences and reactions. Yet many American servicemen and several civilians said at the time or later what it was like to fight in the Great War and how the war affected them. They compiled official records, wrote unit histories, kept diaries, sent letters home, and wrote novels, memoirs, poems, and short stories. Their published medical records, including case histories of psychiatric casualties, contain vivid evidence of the impact of combat during the war and afterward. Several thousand AEF soldiers who lived into the 1970s responded to the World War I Survey, a questionnaire that the U.S. Army Military History Institute sent out to surviving veterans.

We have to use this information carefully. Postwar reconstructions of what happened were subject to distortions of memory and reflected not simply immediate wartime experiences but later thoughts and occurrences as well. Some of the memoirs so closely resemble famous war novels and films that it is hard to say whether the veterans used works like *All Quiet on the Western Front* (itself an imaginative reconstruction of the war) as models or whether those books mirrored with remarkable accuracy the lives of World War I soldiers or both. Several battle narratives tell what the author thought he should have felt and done in battle as well as what really occurred. All are shaped by the writer's location when events described took place, which restricted what he could personally witness and colored his recollection.

Soldiers sometimes exaggerated and often chose not to tell what they knew. They realized that censors read their letters and wrote accordingly; and because they wanted to reassure their loved ones (and probably themselves), their letters sometimes made the war less malevolent than it was. Some could not bring themselves to say what they had experienced and wrote about superficial details or said nothing. Finally, those who chose to record their experiences or answer the World War I Survey questionnaire may have felt differently about the war than those who kept silent. Still, the evidence American soldiers and veterans have left tells us much.

It discloses, to begin with, that during the war very little happened to many of them much of the time. For this information, we can thank the compulsive diarists who wrote about monotonous days filled with trivial events, repetitive training, and mindless labor. Soldiers recalled, for instance, how they were made to hurry from place to place for reasons that were not very clear, then waited endlessly for something to occur. Even in the trenches, men spent days, even weeks, when nothing exciting took place.

We also learn that many American servicemen who stayed behind the lines, doing work essential to a great army of those years—administering, providing supplies, compiling records, and planning operations—lived in their spare time like tourists, with the United States government paying for their holiday. One of these soldier-tourists was Lyman E. Pratt of the Twenty-sixth Division. During the summer of 1918, while the Twenty-sixth was engaged in brutal combat, Pratt rose at 7:00 or even 8:30 in the morning, went to division headquarters, and worked on payrolls and ammunition orders. He spent his leisure time singing; watching Charlie Chaplin comedies and other films; playing baseball; running races; attending concerts, a boxing match, and performances by visiting entertainers; eating steak and french-fried potatoes accompanied by champagne in a nearby village; and escaping to Paris for relaxation. During this period, Pratt quickly rose from private to sergeant. Of the château where he was billeted, he wrote: "Some class, not costing us a cent. . . . This place seems just as though we were on a vacation in a country town back in the U.S. Located on a hill, great scenery." He also watched shelling at a distance, felt sorry for the wounded he saw streaming into a dressing station, and had to endure sporadic bombing and gas attacks.

Combat troops imagined what people like Pratt were doing and resented it. At the same time, some men stationed in the rear had mixed feelings about being there instead of in the front lines. Captain William D. Haselton of the First Division, who had been transferred to headquarters from the front, wrote his family that he was "almost ashamed" that his chances of getting home in one piece were now excellent and told them he would willingly give up his easy job and comfortable surroundings and go back into the line with a smile on his face. Benjamin Heath, sergeant major in the 325th Infantry Regiment, wrote his mother that he felt like a slacker and that she had a better idea of the war than he did because she, at least, watched movies about the fighting. Still, he did not volunteer for the front but stayed in his office recording casualty statistics. "I am . . . playing little Jack Horner in my own little corner," he wrote, "hoping that the boys that are in the fighting will show what the Americans can really do."

Whether they felt guilty or not or were stung by the jealousy of their comrades at the front, a fair number of the Americans who helped fight from the rear enjoyed many pleasures. Captain Haselton described the day he accompanied Major General Robert L. Bullard, commander of the First Division, to a French army headquarters. As Bullard stepped out of his Cadillac limousine, a "wonderful" band played a French military march, and amid much saluting and handshaking, the party walked to the French army château. French officers welcomed Haselton so warmly he "felt like a king." But the high point was the banquet that followed, "a perfectly magnificent meal, typically French and the equal or superior to anything I have *ever* eaten." Haselton especially remembered the perfume of the food. "If I was only a poet or a musician," he wrote, "I could write a poem or a piece of music that would make that banquet famous in the world's history."

For Major Hugh W. Ogden, Harvard LL.B., a partner of the Boston firm of Whipple, Sears and Ogden, and now judge advocate of the Forty-second Division, duty in France was a series of almost uninterrupted delights; at least that was how he described it to his family. "I am simply having the time of my life," he wrote.

> Up here in this sector where we are battling in mortal combat they don't shell the Staff, or drop bombs on them, or blow up the Division Headquarters, or snipe the Staff officers. . . . We don't bomb their headquarters, and they don't bomb ours. We don't shell their Judge Advocate when he pursues his peaceful round of duty, and they don't shell ours. In other words, we fight like gentlemen up here in this part of the country, and everyone is happy.

Ogden took particular pleasure in the social order at division head-quarters. The army gave him a sergeant major—a graduate of Yale and Harvard Law School and the grandson of Cornelius Vanderbilt—who served him effectively and with deference suited to their respective ranks. When it issued Ogden a horse with full equipment, his cup, he said, brimmed over: "It was a proud day for the J.A. when I went clattering down the main street of the city followed by my orderly at a proper distance, and so on into the country. It was simply grand."

Like Captain Haselton, the judge advocate enjoyed his share of elegant meals in splendid surroundings and wrote of one fine luncheon: "We had hors d'oeuvres, fish, chops, beans, cheese, chocolate pudding, coffee, cigars and a little champagne. . . . a very recherché affair." The company appealed to him as much as, if not more than, the food. He described how he sat at the general's mess and discussed military problems and world politics with "the finest type of gentlemen"—distinguished regular officers like Colonel Douglas MacArthur, the division's chief of staff, and Major General Charles T. Menoher, its commander, and men from Ivy League universities and fine professional schools. "Altogether a very high grade combination. The talk is really first class, not a word of money. . . . It is queer to match it up with some dinners I have eaten at clubs where my cigars, my automobiles, my yacht, my horses, my wife's furs and diamonds furnish the main topic of conversation."

For a while, Ogden became the Forty-second Division inspector and had to face shelling near the front lines. Still, he always saw the bright side. "Of course you know war is Hell," he wrote, "and all of us who are undergoing its horrors over here are heroes. . . . At the same time, and strictly between us and not for publication, lest it detract from the hero business, there are times when we have a perfectly wonderful day. Today is one of them."

The French countryside—the parts undamaged by war—delighted Ogden and Haselton and other members of the AEF. "The country is green and covered with flowers," one soldier wrote in mid-April 1918. "We found France a veritable garden land," wrote another. Even at the edge of the fighting zone were thrilling and beautiful sights: aircraft diving and turning in dogfights, and fighters attacking huge observation balloons. The battleground had its own stark beauty, seen from the proper perspective and at the right time, like the night when Sergeant Cyril B. Mosher watched under a black sky as artillery lit the horizon with flashes that changed from pink to flaring red. "Even the trenches can be beautiful," wrote Lieutenant Quincy Sharpe Mills in June from Lorraine, "when they are trimmed with flowers, and the barbed wire forms a trellis for rambling vines, and shelter for innumerable thrushes and other songsters. . . ."

Among soldiers headed toward the battleground, pleasures encountered along the way mingled with apprehension and distress, as Corporal Harold W. Pierce, a scout and rifleman with the Twenty-eighth Division, discovered during his journey to the front.

Pierce was riding from Paris on Bastille Day, 1918, toward his first battle. At Coulommiers his division passed through a small park where several women were smiling and waving at the young soldiers who had come from America to protect them from the Germans. "Three of the prettiest are standing alone," he wrote.

and as we pass we yell at them and they smile and wave at us. A French soldier walked up to one, grabs her skirts and pulls them up, revealing a perfect pair of feminine legs. Only a glimpse and plenty left to the imagination. . . . She blushes and is embarrassed as the others kid her and laugh. . . . The girl's legs fresh in our minds remind us of other women back home, good and bad, but sweet and feminine.

That night, strange thoughts passed through Pierce's mind, "war and its horrors, legs, cheering crowds, legs, refugees, more legs. . . ."

When they reached their destination, the Americans encountered hardships that few had endured until that time. At the front, their basic needs went unsatisfied or were appeased just enough to keep them fighting. They sheltered themselves in water-soaked foxholes and cold muddy trenches, covering their bodies with filthy, tattered clothes and mud-caked blankets. For days at a time, they lived without enough to eat. "To be in the front line of the American army at that time was to go hungry," wrote Lieutenant Hervey Allen. Letters and diaries of other men record how they craved food all the time and how monotonous their rations were. Troops in combat suffered from an almost unslakable "battle thirst." Private William A. Francis of the Fifth Marine Regiment told how at the Battle of Belleau Wood, after a water detail failed to appear, "we felt as though we would go mad for want of drink. I started digging as fast as I could and came to some wet clay. I put this to my tongue. It helped a little, as it was very cool."

There was plenty of water in ditches and shell holes—along with poison gas, decaying bodies and body parts, blood, and human and animal waste. Nevertheless, soldiers drank it and took the consequence, including dysentery, or blinded themselves by rubbing mustard gas–tainted water into their eyes. The army had taught them how to dig slit trenches and build latrines and had given them equipment to purify drinking water; but some men either could not or would not follow correct procedures. Private Wilder Hopkins of the Thirty-second Division wrote that, "at the battlefront there was no such thing as sanitation." Some men relieved themselves wherever they happened to be. Soldiers lived in filth continuously, sometimes going for months without taking off their breeches.

Vermin thrived in these surroundings. For weeks the Americans lived with lice crawling on their bodies. A soldier told a reporter for the *National Geographic* that he did not mind guard duty because the cold nights kept the lice quiet; but he dreaded the coming of day "when I must crawl back into my dugout and try to sleep and know that I shall have to lie awake and feel 'them' crawl." Private First Class Walfred L. Walker saw rats as large as half-grown opossums falling through the tar paper in the ceiling of his dugout, landing on soldiers, and then running like frightened rabbits. It was hard for him to keep them from getting under his blanket. Sergeant Mosher told his mother he had gotten used to rats dropping on his blankets and playing around his feet. But when he woke up suddenly and found himself knocking one off his neck, it was "something else." Sergeant J. Walter Strauss, asleep in a pup tent on an old battlefield in Belgium, woke with a start. A rat was chewing on his hair. "It was tough sleeping after that," Strauss said, "and I became exhausted from lack of sleep."

Sleeplessness, days and nights at a time, was the lot of front-line troops. Those who got used to carnivorous insects and omnipresent rats and could sleep jammed next to one another in dugouts and foxholes or doze off near-frozen in soaking uniforms were wakened by gas alarms, some of them for real gas attacks. In Sergeant Charles R. Blatt's unit, about a dozen gas alarms sounded between sunset and daybreak, forcing everyone to put on uncomfortable masks. Repeating nightmares, in which he felt unable to move, terrified Corporal Pierce and made him afraid to doze off. Many men saw the horrors of the battlefield reenact themselves in silent dreams. Even a trip to recuperate behind the lines did not guarantee rest. When the Twenty-ninth Division was taken out of action for a time, 2nd Lieutenant Joseph D. Lawrence was assigned to a bed with another officer, but each man smelled so badly that they both found it hard to fall asleep.

Going without sleep, marching long distances with a heavy pack over roads that sometimes turned to liquid mud, and the exertion of battle wore troops out physically and mentally. According to George C. Marshall, a colonel on the AEF staff, during the last offensive of the war many American soldiers died of exhaustion.

What they observed on the battlefield troubled many of these men, particularly those new to battle. They fought on a gloomy landscape with shattered stumps of trees and ruined buildings and ground so torn up that they could hardly associate it with the earth they knew. Everywhere they saw bodies of men and animals, blackened, maggot-covered objects. The sights made Corporal Vaughn E. Timmins vomit. Wilder C. Hopkins, a teenaged private, responded in a clinical way, taking careful note of the shapes and positions of the dead: "In one place a man's head was lying with none of the body anywhere in sight. Another part . . . with all of the facial features remaining but the center of and back of the head completely gone as was the body." After a platoon in Corporal Ralph T. Moan's company attacked some Germans with machine guns and grenades, Moan noted in his diary that one of the Germans had his head blown off. "[I]t made a ghastly sight, suspended in the barbed wire." A shell landed not far from Lieutenant Lawrence, who had to walk carefully to avoid stepping into "a bloody mess of flesh and scraps of an American uniform."

At Château-Thierry, German gunners made several direct hits on a trench near Hervey Allen. The next day, Lieutenant Allen and another officer took a detail to the trench and, using blankets and shelter halves, picked up hands, arms, and other parts and buried them. The explosions had smothered some of the occupants and shredded their bodies, which fell apart as Allen's men pulled them out of the dirt. Later, Allen came upon the wreck of a downed airplane. The aviator, whose buttons identified him as an American, was still sitting in his seat. He had been burned to death and there was nothing left of him but a "blackened egg-shaped mass."

Troops saw comrades and enemy soldiers grievously wounded. During the Meuse-Argonne offensive, a German artillery barrage tore through Harold H. Wadleigh's unit, part of the Eighty-ninth Division. "My platoon unlucky," Wadleigh recorded in his diary. "We lose heavily." He described how a private in his unit tried to run with both legs off at the knees. In the same battle, Wadleigh observed a German soldier in the same condition begging the Americans to kill him.

Distance sometimes made it possible to watch the most horrifying sights with detached fascination. Second Lieutenant Louis F. Ranlett observed two columns moving up a hill half a mile away. A fountain of smoke rose into the air between them, and after it cleared, one figure in each column struggled to stand up and staggered away; two other figures raised their shoulders and "fought like mashed ants" to free their shattered legs. Most of the figures looked like dead rags, and some had vanished altogether. Corporal Pierce watched through field glasses as a battalion of green American troops with inexperienced officers moved across a field about two miles away. Pierce realized that many of them would be dead in a few minutes when enemy artillery caught them in the open, but he could not stop watching. He felt like a bird hypnotized by a snake. As the troops reached the middle of the field in perfect order, he heard a distant "crumph" from an enemy battery, followed by a long whistle. Then he observed the shells blowing them apart.

Some battle noises, like the sound of friendly artillery, were reassuring to those who learned to distinguish them from hostile sounds. Lieutenant Ranlett recalled how an Allied barrage seemed to shield him as he prepared to move across no-man's-land: "The sound of the shells passing overhead formed a solid, invisible dome. The sound filled the air. It seemed as though one could reach out and touch sound, pull it away from one's head, butt one's helmet against it." But troops in battle were usually disturbed and frequently terrified by what they heard—whining and cracking machine-gun and rifle fire; screaming, screeching, and freight-train sounds of large shells; the head-splitting concussions of bombs and artillery rounds; and ominous noises of lower amplitude: the dull thud of distant mortars; the swishing, humming, and whistling of large projectiles; the "flop," "flop" of poison gas shells; and the screech of a Klaxon sounding a gas alarm. Sometimes, after a salvo detonated, they heard screams of such intense agony that it seemed no human being could make them—sounds impossible to rid from one's brain—and the cries of men begging to be shot to death. Shrieks and moans of the wounded tore at the feelings of their friends who could not try to rescue them without becoming targets for waiting snipers. For Lieutenant Allen, some of the most terrible sounds of battle were faint noises in caved-in trenches where explosions had buried men alive.

The acrid scent of exploding shells and charred buildings; the odors of poison gas, of muddy ground, of excrement; and the sweetish smell of corpses that pervaded the battleground affected AEF troops deeply and lastingly. Corporal Pierce remembered traveling for two miles over a recent battle site that reeked of decaying flesh. Several days after a battle, Private First Class Thurmond Baccus of the Eighty-second Division wrote from an area where the burial squad had still not finished its work: "I had rather smell gas than odor of men and horses." The men in burial details lived with the stench of the dead, which permeated their clothing and stayed with them when they went to eat and sleep. Major Raymond B. Austin of the First Division sent his men out to bury dead French Moroccans lying near his command post "or else be almost driven out ourselves. Sights don't trouble me, but the other—no one ever gets used to that."

Facing fire was something else most men never got used to—the experience of hugging the forward inside slope of a foxhole while bullets buried themselves in the side behind; walking or running while clouds of metal whistled through the air; being bombed or shelled. "To be shelled is the worse thing in the world," Hervey Allen declared

> It is impossible to adequately imagine it. In absolute darkness we simply lay and trembled from sheer nerve tension. There is a faraway moan that grows to a scream and then a roar like a train, followed by a groundshaking smash and a diabolical red light.

Lieutenant Lawrence recalled how he gritted his teeth and clenched his hands and drew his muscles rigid while shells exploded near his foxhole, and how when a large shell screamed a few feet over the heads of his men, they fell to their knees as if they were one man, throwing the column into disorder. Lawrence noticed that even veterans jerked and twisted as they lay under a heavy barrage. Some men began to cry. During his first bombardment, Lieutenant Ranlett's whole body shook convulsively, as if he had a terrible chill, his knees shook, his fingers and hands moved involuntarily, and he noticed that his voice had a strained, high-pitched sound. Describing how it felt to be in a dugout during a German bombardment Corporal Pierce recalled:

> I am soon a nervous wreck. I lose control as the bombardment wears on into hours. . . . I want to scream and run and throw myself. My gas mask irritates me and I am on the verge of throwing it off, gas or no gas. My throat is dry and cracked from the mask but the saliva runs from my mouth and swishes around on my face. When I hear the whistle of an approaching shell I dig my toes into the ground and push on the walls of the dugout.

Under bombing and artillery attack, troops felt utterly helpless, incapable of responding to what was being done to them, uncertain, paralyzed, afraid to perform basic physical functions. Ralph Siefert, a sergeant with the 103rd Sanitary Train, had gone behind a stone wall to relieve himself when he heard a shrieking sound directly overhead, and almost in the same instant, a bang. "I knew it was a shell"; he wrote his father. "It scared me, so I used all my paper in one wipe and grabbed a handful from a pile that was along side of me, used it and pulling my britches up, made for a dugout. . . ."

An emotion that battle commonly evoked in these men, even in the brave, the stoical, and those eager to fight, was fear, which took many forms. Before their first battle, new men not only worried about combat itself but were afraid they would succumb to fright. Veterans feared crippling and disfigurement. Lieutenant Allen remembered how he heard someone playing a sentimental song as he was getting ready to go into battle and how the music aroused in him fear and nostalgia at the same time, an acute longing for his loved ones, despair, and anxiety about the next day's "indignities," together with a sense that the war would last forever. He called all this the "Just-Before-the-Battle-Mother" feeling and noticed that others felt it too. It paralyzed him mentally and physically until "the great machine of the army" laid its "iron touch" on his mind and body and enabled him to go on. Before his first battle, Corporal Pierce heard a band in a nearby valley playing taps for the dead of its regiment. The music left him with a "helpless, hopeless," frightened feeling.

In much of the American army, a strong fighting spirit prevailed; yet even the freshest, most aggressive troops felt anxious about entering combat. Before his first battle, marine private Malcolm D. Aitken felt an uncanny fear that he believed only somebody who had been in the lines could understand. E. G. Melvine, a sergeant from Tennessee, compared his apprehension, after being told he would go over the top the next morning, to the feeling of a man condemned to die. All his buddies, he recalled, thought the same way. Sergeant Blatt felt certain that if he rose from his trench he would not live for an instant; but he got up anyway, controlling his panic. Just before an attack was about to start, one marine told his buddy he dreaded to go because he was afraid he would not come out alive. An instant later he was killed.

Few of these men, perhaps none of them, had ever experienced the kind of terror they felt on the battleground. Ralph Moan felt "drunk with fear" as he ran messages through sweeping barrages that reminded him of enormous windshield wipers. Louis Ranlett was "scared into jelly" by his first German artillery bombardment. William Langer recalled that, with the earth quivering and huge chunks of clay and stone flying in all directions, "we were all scared, just as scared as human beings can be short of complete breakdown."

While some men recalled in postwar accounts that they had not been afraid in battle, or that their early fears went away, others believed that anyone who claimed he was not afraid in combat was lying. Even soldiers who did not appear frightened admitted later that they had been hiding their terror. When Sergeant Blatt rose from his trench, he made "a show of camouflaging my panic, by yelling encouragement to the others below."

Sometimes fear became impossible to hide or control. As they advanced toward the enemy, soldiers trembled and jerked, vomited, whimpered like children, coughed from fear, or relieved themselves involuntarily. Blatt recalled men of his unit "chalk white with fear" as they hurried, doubled over, through artillery fire. During the Meuse-Argonne battle, when an officer ordered another sergeant to take his platoon and clean out a German machine-gun nest, the sergeant's teeth chattered so hard that his helmet banged rhythmically on his head.

Besides anxiety and fear, combat evoked a mixture of other emotions and thoughts in AEF troops. Some imagined the battle they were fighting would go on forever. Some felt an intense desire to get it over with as quickly as possible, even hoping that death would take them out of it. Others concentrated on surviving. Many became "indescribably sad" as they saw friends die. Survivors recalled wondering if their number was on the next shell or bullet. A good many thought about their mothers. Private Harvey C. Maness recalled how he had been "praying to God for quick victory & a nice dry warm bed [and] a nice mademoiselle to nestle up to" and how he had thought of home and loved ones, safety and death.

Some remembered feeling excited, composed, detached, or all of these at once. Lloyd Foster of the Thirty-second Division wrote shortly after the

Aisne-Marne offensive that it certainly had been "an exciting experience to march right into the face of machine gun fire." He had been as cool, he said, as if he were in an ice-cream parlor, and his buddies had all seemed quite calm. Edward A. Davies, an infantryman in the Seventy-ninth Division noticed that machine-gun bullets were falling around him like hail. "It was a very peculiar sensation," he recalled, "to have these little steel pills flying around, knowing that one of them could kill you if it hit you, but strange to say, after a while you didn't pay any attention to them."

In the frenzy of battle, men acted in odd ways. Lieutenant Lawrence recorded several incidents of irrational behavior, including the time, during a heavy shelling, when a distraught sergeant attacked him; and one occasion when a terrified fellow officer, imagining that Lawrence was a German, nearly shot him. Lawrence saw one man cocking and firing his rifle over and over without bullets. At another time, while his men were being bombarded, he watched an eighteen-year-old private draw himself into a knot, trembling all over. This so upset a corporal that he got up on the breastworks with his back to the Germans and began cursing the young man. The corporal was not hit. Lawrence also recalled how two officers and a sergeant ran up to his company as he was forming them into line, the sergeant waving his pistol in the air and shouting "Don't get excited, don't get excited."

To American troops, firing at enemy soldiers could seem like target practice or hunting. Sometimes it turned into massacre. "How we did cut the Germans down when they tried to cross the wheat field," Private Francis recalled about the battle at Belleau Wood. "The wheat was just high enough to make good shooting, and when we hit one he would jump into the air like a rabbit and fall." Even wounded men in no condition to attack the Americans and men who had surrendered were sometimes shot or bayoneted. Francis reported that the marines were sent to Belleau Wood with orders to take no prisoners. While some were too squeamish or too preoccupied to kill all the Germans they saw and instead took hundreds captive, Francis noticed other comrades shooting enemy soldiers as they raised their arms to surrender. "We are not men any more," Edward Davies remarked after a few days in the Meuse-Argonne battle, "just savage beasts."

Despite such incidents and the ferocity of combat, front-line troops of the AEF developed mixed feelings about the men they fought. Many of the Americans, including some who hated the Germans most bitterly and dehumanized them, admired their skill as fighters. Some even came to identify with their enemies as fellow suffering humans and sensed their pain. Respondents to the Army Military History Institute questionnaire commented repeatedly that the Germans were "human beings" and "just like "us." Griffith Adams, a private first class in the 105th Machine Gun Battalion, wrote his cousin about the way his buddies rushed out when German prisoners came by, taunting them with comments like "Where is the Kaiser?" and "This is your march to Paris." Then he added: "We really feel sorry for the poor fellows, and such young ones too, but they would do worse, if it were just the other way." Lieutenant Lawrence was horrified by the screams of pain he heard when he and his men fired into a group of Germans hidden in thick underbrush. Corporal Pierce described an action in which soldiers of the American Twenty-eighth Division caught a group of Germans in an open field along the Vesle River and killed them: "The men that shot them are not proud of their work and wish it could be avoided. In the heat of battle men do not realize that the enemy is only a scared, frightened boy like we are, killing for self preservation and because he has to and hating it as bad as we do."

When Sergeant Blatt, another sergeant, and an infantry lieutenant went to reconnoiter in the Argonne Forest, they came upon the body of a dead German. A short while afterward, Blatt wrote that as the three of them stood there looking at the corpse,

> A most surprising thing occurred. The infantry officer broke the silence. "It's a damned shame, isn't it?" Unmoved by the sight of our own dead, the German seemed to bring to all three together a realization of the futility of war as expressed in the killing of human beings. And these Germans didn't look like such bloody monsters. Quite an ordinary everyday crowd.

While thousands of AEF infantry and artillery men fired their weapons willingly or without apparent qualms, some of the Americans were reluctant to kill or felt remorse afterwards, at least in their early battles. A soldier in the 168th Regiment observed: "We felt rather queer and empty at the stomach when we actually realized for the first time that there was a need to kill or be killed. The idea of dispatching a man just because you saw him first was not refreshing." The first time Harold Pierce was ordered to shoot, he could not see anything to fire at and enjoyed the kick of his rifle as he squeezed the trigger. But a few days later when he tried sniping at some visible Germans, he missed them twice, a failure he attributed to thoughts about the senselessness of what he was doing. Though he pretended to be upset at his poor shooting, he was glad the targets escaped. Sergeant Norman L. Summers of the Forty-second Division reported how a machine gunner, who was credited with killing a whole company of Germans during the fighting on July 14-15, became so sick of the slaughter that he had to be carried from his gun. After a

German counterattack in Fismette, Pierce met an American who claimed he had killed twenty Germans but was so ashamed of himself that he stopped firing. Decades afterward, former Private First Class Henry Fritscher prayed each night for the people he killed in the First World War, consoling himself by surmising that they were better off dead.

Special circumstances led American soldiers, at times, to relent from killing. Private First Class William H. Houghton of the Sixth Engineers was captured by the Germans who assigned a young boy, perhaps fifteen years old, to escort him to their headquarters. The boy, who had what Houghton remembered as "a grin a mile long," seemed poorly trained. He had failed to notice that his prisoner had a large jackknife hidden in his uniform. Later Houghton wondered why he had not used it. There were many possible reasons, considering he was behind enemy lines. But the actual reason, he said, was that he did not have the heart to kill a young boy in cold blood. Years later, he said he slept better because he had failed to use that knife.

American troops handled the stress of killing and surviving in various ways. Many, as we will show in the next chapter, were sustained by bonds of comradeship, by personal and family honor, or by patriotism and belief in the rightness of the war. Some just concentrated on doing their jobs and trying not to think, keeping busy giving or carrying out commands. Officers and noncoms focused on moving their men and directing fire—tasks they had been trained for in the preceding months or years—getting a chance to perform what they had only heard of in lectures or read about in manuals. Major Austin, a field artillery officer who had advanced with his guns alongside troops of the First Division, said how interesting it was to get a really close-up view of the effect of artillery fire on an infantry line, the line of Americans where Austin was situated. But generally, these troops were too busy living through each moment of battle to meditate on their circumstances. This was fortunate, Lieutenant Allen observed, "because with time to brood, conditions would often seem intolerable."

Religion helped—doctrine, feeling, and symbols and ritual. Men found deep comfort in religious ceremonies and in the thought that God watched over them. Harold Pierce imagined that Christ stood at his elbow and believed his mother's prayers and faith were keeping him whole. Corporal Pierce felt certain that the New Testament he read from regularly and carried in his shirt pocket during battle would stop shrapnel from reaching his heart. Sergeant Major Vernon C. Mossman recalled that as a seventeen-year-old doughboy in his first battle, he had not feared death because his Protestant religion had taught him to have faith and that life was preordained. One of Lieutenant Lawrence's sergeants struck a bargain with God, promising to give up drinking if the Lord spared him. (The sergeant failed to keep his part of the bargain.) Lieutenant Phillip C. Shoemaker sensed that both the Almighty and the spirit of his late father guided and protected him. Still, Shoemaker took no chances. He went to mass in a French church and carried medals and rosaries at all times to keep on good terms with the Lord. The closer and faster the shells and bullets came, the faster and more passionately he prayed. As Captain Lee A. Tierney, a field artillery officer in the Twenty-eighth Division, observed the muzzle flashes of a six-gun enemy battery that was zeroing in on his position, he "prayed like the Moslems at noon."

The men of the AEF had a knack for seeing humor in their situations, and this helped them at the front. Sometimes they only affected an amused attitude toward what terrified them or pretended to be cheerful when they dreaded what awaited them. Doughboys joked about their boring, awful tasting food (canned beef was "monkey meat," corned beef was "canned willie") and the vermin with whom they cohabited (lice became "cooties," "pants rabbits," and "seam squirrels"). They told grim jokes in the midst of artillery attacks and made fun of the horrifying (ambulances were "meat wagons"). They laughed at sights, like headless bodies, that made them nervous. Private Francis could not stop chuckling as he watched another soldier taking the glasses of a German intelligence officer as a souvenir while the officer, who had been shot repeatedly, still remained alive.

Some of them witnessed absurd sights in the midst of battle—like two drunken doughboys staggering down a street during the Argonne battle in full dress suits and top hats while shells landed around them. J. Walter Strauss recalled the night when his unit was out preparing barbed wire and a German flare went off, lighting the battlefield with a terrific glare. Everyone froze, some sitting, some standing, some holding the wire or a shovel, and others perfectly still in some other position. "It really was a funny sight, and we howled laughing afterwards! But when we were frozen and waiting for a machine-gun bullet to finish us off, it was not very funny!!" During the Argonne battle, Edmund P. Arpin terrified a German soldier by suddenly emerging from cover, aiming his pistol, and pulling the trigger. In the excitement, Arpin had forgotten to load his weapon; but the German let out a wild scream, half-somersaulted backwards, and dove down a hillside, yelling as he escaped. Arpin and a sergeant who was with him rolled on the ground with mirth. Horatio Rogers was out with a detail stringing wire on no-man's-land when shells began to detonate nearby. Everyone in the group wet his pants, and when they returned, they were all weak from laughing.

Some doughboys used drink—wine, beer, and distilled spirits—to cope with combat, though they usually had to obtain it on their own since the American army did not issue alcoholic rations before battle. (The British gave rum to American troops in their sector.) Nicotine was the drug of choice—as tranquilizer and stimulant—although the War Department prescribed that it was to be used in moderation. Observers noted the power of tobacco. Lieutenant Frank P. Isensee watched officers and men of the Twenty-ninth Division leaving their jump-off point during the Argonne offensive smoking cigars and cigarettes and shouting commands like "keep that line straight" and "get your right direction." Major Austin noticed how American First Division troops kept on advancing while shells landed among them, most of the men smoking cigarettes; all were calm, not talking much.

As they continued to experience battle, soldiers began to change. The effects could be seen in men who had gone into battle determined or even exuberant and came out exhausted, their bodies stooped as though they seemed to be cringing from imagined artillery shells, their faces gray, and their eyes heavy-lidded and staring. Part of the inward transformation of these men was a new way of feeling about wounds and death. Sergeant Blatt wrote that thoughts of death or injury began to hold fewer terrors. "We were too thoroughly accustomed to seeing others getting it, and to the possibility of getting it ourselves." While he strongly wished not to die or be wounded, he decided it would be egotistical of him, with so many falling around him, to imagine that his own death or injury had any meaning. Lieutenant Davies recorded, after several grueling days, "There is no fear in me now. I would go forward willingly and play the game." Observing that the dead men on the battlefield were literally returning to the earth, Hervey Allen began to think of death as simply a biological phenomenon.

Many soldiers, after seeing so many men killed or wounded, developed a sense that living or dying was out of their hands: part of them hoped to survive; another part became utterly fatalistic, thinking themselves already dead. Some hoped that death would be sudden when it came. Some became profoundly despondent. Corporal Pierce told of soldiers who hoped for death as an escape from the war, and Private Francis recalled that he prayed during the Battle of Belleau Wood for the Germans to come into his trench "and either kill us or . . . we would kill them, for the attacks we went through were worse than death." Other men believed the chances of battle favored them or imagined that, in spite of what they saw, they themselves would survive. Reflecting on how he had kept going at Soissons and Belleau Wood, Lieutenant Samuel Meek remarked: "You knew there'd be heavy casualties, that the odds were against you. I guess each man just figured it wouldn't be him."

Troops became mentally anesthetized, their minds deadened to thoughts of what might happen to them. As one survivor observed, "think-ing is bad, even for soldiers who have schooled themselves to look upon death as the common fate of all." Private Aitken wrote his parents, who wanted to know how he felt in his first battle, "I didn't feel at all." Private Duncan M. Kemerer remembered acting like a robot, without nerves. "We had to develop a numbness and an unfeeling attitude toward it all," Corporal Mahan observed. "Other[wise] we could have lost our minds." Guy Bartlett, a private first class in the Twenty-sixth Division said, "I lived in a dream state." But every now and then, they witnessed something, like the sight of dead comrades strewn on a battlefield, that made the numbness lift and started them feeling and thinking thoughts their minds had pushed away. This happened to Lieutenant Ranlett. After a long period of combat, he went off by himself to find his regiment and passed through a camp of tar-paper shacks where a "hurricane bombardment" had landed on a large number of Germans. "Being alone, being calm, being relatively unhurried," Ranlett "shuddered as I had not done at many equally gruesome sights during the days just past."

Thousands—officers, noncoms, and enlisted men—tried to avoid bat-tle physically. For some, this was a voluntary act. Others found it impossi-ble to continue fighting. A few pleaded not to be sent into combat or wounded themselves, fell out on the way to the line, or, when battle began, refused to leave shelter. Some ran or employed more subtle methods to avoid enemy fire, like going to the rear to pick up ammunition that did not exist, lagging behind and escorting prisoners to collecting areas, or, in the case of officers, directing their units from shelter when they might have done their jobs better in the open. On a few occasions, as we have noted earlier,* masses of troops panicked and left the battle in disorder. Thus in late September, while the Thirty-fifth Division was withdrawing from the battlefield, the Germans caught it in machine-gun and artillery fire, send-ing its soldiers stampeding to the rear. A retreat by part of the Twenty-eighth Division turned into a rout.

When Lieutenant Lawrence's company went over the top on October 10, the first sergeant and several other men ran away. "Many men cannot muster the courage to jump from a trench and rush the enemy in the open," Lawrence wrote, "whether they wear American uniforms or not." Afterwards, he had to patrol his own lines all night to keep his men from escaping. He would force one back, and another would try to slip by. The sergeant's desertion hurt Lawrence's unit very badly. During the third day of the attack, during a German bombardment, American artillery began to shell their own troops in the front lines. The American gunners had the

---

*In the discussion of the Ninety-second Division, chapter 6.

wrong coordinates, but Lawrence's men could not send up a signal rocket to stop them because the sergeant had taken the rockets when he left.

It is hard to know exactly how many American soldiers avoided battle. Lawrence thought that about four or five men deserted out of every company (256 men at full strength). The army convicted only 1,553 men for desertion in 1918 and sentenced 24 to death. (None of these sentences was ever carried out.) Yet these numbers tell only part of the story. Major General Hunter Liggett, commander of the First Army, estimated that during the Battle of the Meuse-Argonne about a hundred thousand men were straggling in the area. General Bullard remembered "the great number of . . . dead-beats we herded up" and recalled how French villages were filled with Americans who should have been at the front. At one point during that battle, the Third Division was able to locate only some four thousand of its official complement of over twenty-eight thousand officers and men, while the Fifth Division could account for only about five thousand of its troops. By the middle of October, the Thirty-second Division had fewer than two thousand armed effectives. Thousands of the missing soldiers had died or been wounded or taken sick; but how many straggled because they were slow or lost and how many were trying to be someplace else when the fighting began is impossible to decide.

The AEF spent considerable effort trying to keep its troops at the front. Military police and special details patrolled rear areas and searched kitchens and aid stations hunting for stragglers. At one point, General Pershing wrote his corps and division commanders to notify young officers orally to shoot their men down if necessary to keep them from deserting, and General Preston Brown, commander of the Third Division, authorized throwing bombs into dugouts if men refused to go over the top. In practice, though, the penalties were usually much less severe. MPs sent men back to the front with signs on their backs reading, "straggler from the front lines" or "deserter." Other punishments ranged from latrine duty, demotion, or removal from command to several years in prison. To prevent the problem from even arising, the army arranged for those who seemed likely to desert or fail in battle to stay away from the front. In the Twenty-seventh Division, troops who would not or could not go were assigned to rear areas.

The great preponderance of AEF combat troops fought valorously, sometimes recklessly. Teilhard de Chardin, a French priest who served as a stretcher bearer near the Americans at Soissons, wrote: "We had the Americans as neighbors and I had a close-up view of them. Everyone says the same: they're first-rate troops, fighting with intense *individual* passion (concentrated on the enemy) and wonderful courage. The only complaint one would make about them is that they don't take sufficient care; they're too apt to get themselves killed." Allied and enemy soldiers marveled at their audacity and the freedom with which they were expended, singing as they went into battle, advancing on heavily fortified positions with little or no artillery preparation, or crawling on their bellies toward the muzzles of machine guns. At times the Americans seemed possessed. Private Aitken recalled the fiendish cries of the Fifth Marines as they charged at the enemy during the Battle of Belleau Wood. During the same battle, marine lieutenant Samuel W. Meek watched a sergeant climb on top of a machine-gun nest and bayonet its occupants before succumbing to gunshot wounds. Private Francis remembered how when the Germans opened up with machine guns, hand and rifle grenades, and trench mortars in Belleau Wood, "we all seemed to go crazy for we gave a yell like a bunch of wild Indians and started down the hill running and cursing in the face of the machine-gun fire," screaming and shooting until reaching the German trenches. Loren Duren of the 58th Infantry Regiment recalled how, as German artillery sent shell after shell screaming at them, his unit struggled toward their objective near the Vesle River "covered with blood and filth," all of them "quite mad . . . screaming with rage," longing to reach the "swine" who were annihilating them.

The troops who straggled or ran and the troops who fought like devils were sometimes one and the same, for soldiers' courage fluctuated from battle to battle, from day to day, and even from one moment to the next. Lieutenant Isensee described how he and other men of the Twenty-ninth Division, who had fought as boldly as anyone could hope for, gathered their courage as they rested. During the Argonne offensive, when Lieutenant Lawrence's men were running out of ammunition, he ordered them to charge a German trench with bayonets. Of about 120 men, only one sergeant, ten privates, and Lawrence leapt out. He called them two more times and on the third try got them to follow him and the sergeant. Then, as the Americans approached, the Germans abandoned their shelter and ran while Lawrence's men coolly fired round after round into their backs.

Some of the bravest people in the AEF felt their courage slip away at times and fantasized about deserting. Lieutenant Ranlett chastised himself for his desire to delay behind the lines while helping a wounded man to safety. "I hadn't meant to shirk," he wrote, "but I had been glad of an excuse. I must get over that." At one point in the Argonne attack, Lieutenant Lawrence, suffering from an infected foot that made it hard for him to walk, was sharing a foxhole with a starved and terrified runner who pleaded to be allowed into the company dugout for five minutes. He told Lawrence it would be like going to heaven. Lawrence felt like saying, "Let's both go back."

Among the ways to go back honorably were to receive a wound, to be poisoned by gas, or to undergo a psychological trauma called shell shock (discussed in a subsequent chapter). Some men hoped for a "blighty," British army slang for a wound that sent you back home alive. In the Argonne offensive, a private who had been wounded earlier and decorated for bravery was pleased when a shell fragment put him permanently out of action. "I was happy to be hit again," he wrote, "because life in the trenches, plugging through the mud and water up to the waist, sleeping in wet damp dugouts is unspeakable." During the battle for Fismette, Corporal Pierce lay in a hole on the field with his brother Hugh, secretly wishing Hugh would be wounded, not seriously but enough to send him to safety. "As for me I would welcome a wound, even a foot or a hand off, just so I am not going to be horribly mangled. . . . I can understand now why a man is so happy when a bullet hits him in the arm or leg. A blighty is about the best a doughboy can look for in this war." A bullet in the leg jolted Lieutenant Edmund Arpin out of a fatalistic depression, but he was less shocked than pleasantly surprised by "miraculously getting off so easy." Taken to an evacuation hospital, relieved of his lice-ridden clothes, placed between clean sheets, and no longer soaked or frozen, he stretched out on his cot "actually giggling with joy." Corporal Moan wrote afterwards how it felt to be saved by a wound that he suffered while running through artillery fire:

> God!
> My head—torn from my body!
>
> Is this what being dead is like,
> Peace quiet, clean white sheets?
> My head must still be here,
>   It hurts me.
> These must be my fingers
>   I can move them.
> And now a voice,
>   "My boy. for you
>   The War is over."
> You hear that, Mom?
> I'm still alive, I did not die.
>   "I'M COMING HOME!"

While some American wounded cried out in pain, others acted quite stoically. Teilhard de Chardin noticed that when doughboys were hit, those who could walk made their way back "holding themselves upright, almost stiff, impassive, and uncomplaining." It was "heart-rending" for Corporal Pratt to see the wounded men streaming into an aid station; but he noted that their spirits remained high and that many of them were laughing and joking. Dale Russell, a driver with Ambulance Section 588, wrote his parents how painful it felt to him to drive his patients on the rough roads from the front. German prisoners "hollered their heads off" all the way to the field hospital; the French would yell once in a while at a jolt, but wounded Americans never let out a whimper. In a ward at the base hospital in Langres, when the "agony cart" came around and the doctor and nurses changed the dressings on exquisitely painful wounds, soldiers tried not to show how they felt. "I have not struck anything so painful in my life," wrote one of them, a man with a leg wound about ten-by-five inches and almost to the bone. "One must not cry out or make a sound during the dressing or the rest of the men will call you baby. This is a disgrace that is hard to get over."

Experienced doughboys knew how terrible wounding could be. Sometimes the victims felt nothing, at first. While Lieutenant Shoemaker was riding in a truck, German shells started to land alongside. He noticed that he was bleeding a great deal and realized he had been hit. At the end of a day of battle, someone told Sergeant Cutler that he was wounded in the leg. He unwound his puttee and found it filled with blood. Yet in the excitement, he had not noticed it at all. The only pain he felt was from barbed wire. Other men knew immediately when they were hit. Private Milton B. Sweningsen remembered how it felt when shrapnel struck the back of his neck: "I thought my head had been knocked off." Lyle Cole thought a sledgehammer had smashed into his foot. As the Twenty-sixth Division pushed through Belleau Wood, a piece of metal hit Private Roland F. Graves in the collarbone, shattering it and opening his trachea. His blood drained into his lung, and he started to suffocate, but he turned onto his stomach allowing the blood to run out. For hours he lay in a hot wheat field with no water near enough to reach. Then at about eight in the evening, he gathered up enough energy to hike back to a dressing station.

When a shell fragment penetrated Lieutenant Louis Ranlett's left eye, he began to hallucinate. He experienced "a flash of terrific brilliancy: A world of globes of light." For a while, he could neither hear nor feel, but he sensed that he was flinging his arms up to ward something off. Then his knees buckled, and he slumped face down in the trench. There seemed to be two of him, one wounded and weak and the other perfectly all right. The unharmed Ranlett looked at the wounded Ranlett, who was completely blind, and tried to speak and act through him. The sick Ranlett obeyed feebly or not at all, which puzzled and astonished the Ranlett who had not been hurt.

In the Battle of the Meuse-Argonne, Sergeant Blatt was advancing with the 305th Infantry into a belt of artillery and machine-gun fire when something struck him at the hip with "fearful" force. He found himself rolling on the ground and moaning in unbearable pain. Horrified at the idea that a shell had carried away his leg and hip, he consoled himself with the thought that the wound would kill him, but stretcher bearers found him and brought him back over a rough and painful road to an aid station where medics put on a dressing. From there he was taken to another station where a large crowd of walking wounded had arrived, and he saw litters covering the ground. His next stop was a field hospital. Then an ambulance picked him up and took him on a "punishing" ride to Red Cross Evacuation Hospital no. 110 for surgery.

The sergeant found himself in the "shock" ward, though he thought it might be better named the chamber of horrors with its "constant flow in and out of the operating room of desperately wounded men, the scream- ing when dressings were changed on the stump of an arm or a leg recently amputated, a head gashed up, part of a face blown away, or a stomach punctured by a dozen pieces of shrapnel; the insane gibbering, mouthing and scorching profanity of men partially under ether." Eventually, Blatt left for American Base Hospital no. 1 in Vichy, where after about seven uncomfortable weeks he could stand up for nearly a minute. He returned to America in February 1919, had more surgery, and was mustered out May 23, 1919, with a marked limp and little ability to walk without a cane. At the end of June, he had a third operation at an army hospital.

While Sergeant Blatt was injured by flying metal, poison gas hurt other men as seriously or even more severely, and to some men, it was even more frightening. Private Elmer C. Goodrick wrote his sister that he feared gas more than shells, "for it is goodnight if you are caught without a mask." It is easy to understand why Goodrick felt that way. At Cantigny, in one of the earliest American operations, gas shells landed on the Eigh- teenth Infantry Regiment. Troops found it difficult to breathe. Some noticed that everything was starting to spin. Vomiting, coughing, with lobster-red-swellings in armpits and scrotums, their eyelids sticking to- gether, the casualties were evacuated in long lines, each man holding the shoulder of the man in front of him.

Medical care varied greatly. It sometimes took hours or days to get wounded men proper help, and there was always the danger that German shells would land among them after they had been collected for treatment. Surgeon Harvey Cushing reported seeing casualties who had been lying out all night at a railway station untouched, waiting for the train. A large number were sent, soaked, to a hospital nearby where the surgeons worked all night on sucking chest wounds, amputations of the thigh, muti- lations, and other wounds and operations. He remembered only two wounds that were not "stinking." Yet doughboys also reported how skillful army doctors were and on the kindly treatment they received. Some sol- diers even enjoyed their time in the hospital. Private Raymond Corkery, who had been accidentally run over by a gun and then had contracted influenza and pneumonia, was sent to a hospital where attractive young Irish nurses took care of him. Norman Summers recalled how Red Cross people gave cigarettes, tobacco, and chocolates to wounded men and how, after he was gassed, a nurse brought him two wonderful slices of bread with blackberry jam. He especially admired the kindness of the American medical attendants and their willingness to help the doughboys. "It does one good," he wrote, "to see how tender the big rough American can be toward the patient." Another soldier described the young man who bandaged his leg wounds at a first-aid station as "gentle as a woman."

As one reads what they wrote about the war, one notices that many of the front-line soldiers developed conflicting feelings about it and about the part in it that they had played. The idea that the experience had been awful appears again and again, but they were glad they had gone through it and would not have missed it for anything. There were several reasons why individual soldiers displayed these varying attitudes. While they remembered hardship and terror, they also recalled the happier parts of their war experience, like the humor of certain situations, their feelings for their comrades, or the pride they took in doing their duty and forcing the enemy back. During the war, a soldier's outlook could fluctuate drastically, going from buoyancy to weary disillusionment and then to spiritual reinflation. For some men, all it took to revive ideals and optimism was a rest behind the lines, a change of clothes and a bath, the sound of a military band, or the sight of little children smiling at the Americans who had come to France to aid them; for others, nothing helped.

Men in battle experienced a heightened perception and an excitement they would never forget. Lieutenant Lawrence remarked about a savage fire fight in which he had blown off the top of a German soldier's head: "it is horrible, but then, while the heat of battle was upon us, it was thrilling." Hervey Allen recalled how alive he felt in one terrifying emergency when excitement "photographed . . . forever" an entire scene on his retina. Describing how he had crossed the Vesle River over a broken bridge swept by machine-gun and rifle fire, Allen remarked: "That was undoubtedly the most intense moment I ever knew. For me, it was the great moment of the war."

Deprivation and the painfulness of living and fighting on the battleground heightened the pleasure of ordinary things like food and water and sex. Just surviving battle produced unforgettable satisfaction. As Sergeant Cole said of the Armistice: no one who had not been through combat, seeing friends killed or wounded, "living with death for weeks at a time, . . . expecting their turn would be next, . . . can ever understand the feeling of relief, the thankfulness, the overpowering love of life, the reprieve to live again with our loved ones, it was wonderful."

While some of these men opened up once the fighting ended, others kept their memories to themselves, except, perhaps, when they talked to one another. Many of them believed, as Sergeant Cole did, that it was impossible to convey what they had experienced to people who had not been at the front. How could civilians understand the young men whose numbness and exhaustion and staring eyes reflected days of battle?

Soldiers of the AEF had other reasons for keeping silent. Some did not want loved ones to share their suffering. Before Corporal Clarence Mahan reached home, he decided to clear his mind as well as he could of all the terrible experiences of the previous two years. "It would not be right," he explained in a memoir, "to make my family and friends sad and uncomfortable by inflicting upon them the horrors in which they had no part." Soldiers reassured the people back in America that everything was all right—even when it was far from all right—and told them things about the battle that anyone in the battle would have thought fantastic; for example, Lieutenant Shoemaker's statement to his mother that "all soldiers killed on the field have a quiet, calm peaceful smile on their faces." Some seemed aware that what they wrote would be disseminated in their hometowns and, perhaps as a matter of patriotism, left out items that might have depressed or horrified their neighbors. Others found the reaction of civilians annoying. Lieutenant Ranlett felt that what silenced the veteran was the listeners who refused to listen but interrupted his recollections with, "It's all too dreadful," or "Doesn't it seem like a terrible dream?" or "How can you think of it?" or "I can't imagine such things."

Yet, some who had fought in the AEF chose not to recount what they had seen to protect themselves. Lieutenant George L. Armitage, a twenty-six-year-old army physician, wrote his parents between battles that his mind was "full of things that I have seen, some of which I will never, never, repeat to a living soul as long as I live. Men who have been here the longest are the most silent on what has gone on and only long for the time when these memories will be drowned out by the love of our dear ones and the pursuit of peaceful duties." Ralph Moan, after relating in his diary how snipers shot about twenty men and how he had seen a man's severed head hanging on barbed wire, stopped and wrote:

> Have decided to cut this diary out right now, for no man wishes after seeing what we have seen to recall them but rather wishes to forget. From now on all we see is HELL
>
> R.T. Moan
> March 30, 1918.

Malcolm Aitken told his brother not to expect him to explain what had happened to him in France because every time he thought of the front he saw friends and comrades falling on all sides. As late as 1935, when he was preparing an edition of his wartime letters to his parents, he still could not bring himself to include some of his experiences. It took fifty years for Duncan Kemerer, severely injured during a battle for the village of Fismette, to tell anyone what he had gone through during the war.

Whether they spoke about it or not, whether or not they understood what had happened to themselves or even realized that something had happened, men like Kemerer and Aitken and thousands of other Americans had been permanently affected by the World War I battleground, by its sights and smells, by killing and fear, by excitement and battle frenzy, and by facing death and surviving in a place where, as Private Baccus said, "love of thy neighbor is forgotten, [and] a man stands before his world his true self—stripped of all the falsities of a sheltered civilization."

11 ||| Motivating the AEF

Although Woodrow Wilson had little knowledge of battle—just a few memories of the wounded men and prisoners he had seen at his father's church during the Civil War—he came to understand, at least in a general way, what the battleground was like. In 1916, while the French were struggling to regain the fortress of Verdun, he remarked about the Great War's "mechanical slaughter" and commented on the "uselessness" of the "sacrifices made." Trench warfare and poison gas had changed battle. By using enough ammunition and expending enough men, it was still possible to capture a strong objective. But where, he asked, was "the glory commensurate with the sacrifice of the millions of men required in modern warfare to carry and defend Verdun?" A few months later, he and his government had to find ways to encourage Americans to sacrifice themselves in France.

Encouraging the Will to Serve

Many Americans needed no inducement to go to war. But to motivate the rest and to build enthusiasm for fighting, the administration used the techniques with which it managed the rest of the war effort: incentives and exhortation, compulsion accompanied by sanctions and threats, and euphemisms and indirect controls.

To fill the ranks, the federal government bombarded men of military age with images—of Uncle Sam pointing at the viewer and telling him to do his duty, or of U-boat victims and children mutilated by German troops. Often the images were of women: the windblown young woman in a sailor suit who says "GEE!! I WISH I WERE A MAN—I'D JOIN THE NAVY"; the pretty teenager wearing a clinging chemise, with her long dark hair blowing behind her, who holds an American flag above her head with one hand and waves with the other toward an army of young doughboys; the Frenchwomen who implore Americans to "Come Across and Help Us"; the liberty goddess who tells young men to "Uphold Our

Honor"; the Belgian woman crucified against her wall with blood dripping down her body; and the drowned mother, sinking with her dead baby through the greenness of the ocean, on a poster with the single word, "Enlist."

Able men who did not heed such messages voluntarily or who were not engaged in essential civilian work entered the armed forces through conscription, a program the Wilson administration began with considerable trepidation. Aware that the draft had led to mob violence when the army had conscripted men to fight in the Civil War, it tried to diffuse and deflect animosity people felt toward government compulsion. Calling its program "selective service" rather than "conscription," it established a network of draft boards comprised of local volunteers so that a man's neighbors rather than the federal government sent him into the army or deferred him. Meanwhile, army officers in the War Department developed the policies that governed those local decisions. As the head of the Selective Service System, Provost Marshal General Enoch H. Crowder observed, draft boards became "buffers between the individual citizen and the Federal Government," lightning rods for hostility that might have interfered with the war effort if it had focused on authorities in Washington.

The president and the secretary of war took steps to encourage Americans to accept the draft. Wilson declared that it was "in no sense a conscription of the unwilling" but rather "selection from a nation which has volunteered in mass." Secretary Baker tried to "create a strong patriotic feeling" in favor of conscription by arranging for local officials and chambers of commerce to popularize it. Baker told Wilson that he was using a "vast number" of agencies throughout the United States to make draft registration a "festival and patriotic occasion." Meanwhile, the Selective Service System announced that it would publish the names of those who registered, enabling local communities to exert pressure on those who had not.

On June 5, 1917, men between ages twenty-one and thirty-one filled out registration forms listing their names, addresses, occupations, ages, and reasons, if any, why they should not be drafted. Each received a small green card with his name and number on it. The numbers were to determine the order in which selectees would be called. Then on July 20 at 9:30 A.M. in a hearing room of the Senate Office Building, Secretary Baker, with a blindfold over his eyes, reached into a large jar filled with thousands of tiny capsules containing registration numbers and pulled one out while cameras clicked and flash powder boomed. The number was 258. Other dignitaries took their turns drawing numbers until 2:16 A.M. the following day.

As the recruits and draftees left for camp, the gaiety and excitement that surrounded them recalled scenes in Paris, London, and Berlin at the beginning of the war when Europeans could not wait for their young men to start killing one another. Communities arranged gala farewells—dances, public meetings, parties with streamers, pretty young women, and sentimental songs. A bonfire of German textbooks scented the air, and a band played "Keep the Home Fires Burning" as a contingent from Hollidaysburg, Pennsylvania, left for camp. In Versailles, Indiana, a father watching his son's train pull away, jumped up and down, waving his hat and crying, "Get the sons of bitches, son, Get the sons of bitches." And as the troop trains rumbled through the towns of America, whistles sounded and bells rang, people cheered and waved the American flag, bands played songs like "Marching through Georgia" or "The Bonnie Blue Flag" and "Over There," young women on station platforms shook hands with recruits through train windows, and Red Cross ladies handed the boys apples and cigarettes, cookies and chewing gum. The atmosphere of these events arose from the patriotism of the American people and from the government's efforts to send enthusiastic troops to France.

The men the government summoned to arms responded in varying ways. Hundreds of thousands rushed to volunteer. Four hundred thousand Guardsmen entered federal service. Draft boards registered some 24,000,000 men and sent 750,000 of them into the armed forces. But from about 2,500,000 to 3,500,000 failed to register, and about 338,000 of those conscripted did not show up for induction or deserted after they arrived at training camps. Some fled to Mexico, others to Canada. In some states, as many as 80 percent of the registrants claimed exemption. Thousands rushed to marry when they learned that single men would be taken first. The neighbors of a Pennsylvania man denounced him after he married his thirty-nine-year-old landlady and claimed that he was supporting her children. Some lied about their health, bribed physicians to exempt them, or feigned medical incapacity—simulating deafness and poor eyesight, swallowing medications that increased their blood pressure or heart rate, pretending to have hemorrhoids, or reporting for physicals wearing trusses for fictitious hernias. Hoping to be turned down as narcotics addicts, a number took morphine or heroin; while others, who preferred prison to the battlefield, simulated burglaries. General Crowder noted that such self-mutilations as removing a finger or a toe or blinding an eye were "rather frequent." Nearly 65,000 filed conscientious objector claims.

Building Morale

The government took various steps to strengthen the morale of those who entered military service. Beginning in October 1917, the War Department issued to every draftee a *Home Reading Course for Citizen-Soldiers*. Much

of this pamphlet was devoted to motivating them, chiefly by promoting
solidarity with the nation and the army and by encouraging them to think
and behave as the army desired.°

The War Department used American history to give draftees a sense
that they were part of a national military tradition and to provide them
with models for behavior. It recounted acts of extraordinary bravery by
Medal of Honor winners and encouraged troops to be cheerful by telling
them how the Army of the Potomac marched home through cold and mud
after the Battle of Fredericksburg "in high humor, the soldiers laughing
and joking at their ill luck." It explained how the government's reasons for
sending them overseas were part of a pattern of benevolent interventions
by Americans—a people with "no taste for warfare and no lust for territo-
ry or power," who had nevertheless fought six major wars for rights and
principles against governments that trampled the rights of others. The
German government was merely the latest aggressor to incur America's
just wrath.

The pamphlet tried to show that the war and military service fit the
egalitarian and individualistic values that Americans were taught. It
claimed that everyone, women as well as men, civilians as well as soldiers,
the rich as well as the poor, served and sacrificed. The army, it explained,
was actually democratic despite its demand for unquestioning obedience,
because in American military service everyone had an equal chance for
promotion and men from every group and region were present in the
ranks. For all its emphasis on discipline, the army was "not merely a big,
soulless machine that moved with mechanical precision" but a "team"
whose members were presumed intelligent and self-reliant.

The *Home Reading Course* said little about actual conditions on the
European battleground, and what it did say was idealized or diluted. Its
bloodless account of combat on the Western Front, "the great game of
war," explained that soldiers were learning to pick up enemy grenades and
throw them back before they exploded. It informed its readers that although they would probably be afraid in battle, they would conquer their
fear.

In its advice to draftees, the army stressed that "nerve and fighting
spirit" did more to ensure victory than the best weapons in the world. On
the drill ground and in the field, it worked hard to instill these qualities in
its men. It developed camaraderie by making them face hardship and
harassment together and engaged them in ceremonies that united them to

°The course's thirty lessons covered such subjects as military courtesy, physical fitness, equipment,
and personal hygiene taught with the assumption that the reader knew next to nothing. For instance, it
instructed soldiers in how to chew their food and how often to brush their teeth and empty their
bowels.

regiment, army, and nation. Music played an essential part in morale building. The Committee on Training Camp Activities sent singing coaches to the camps to teach the men patriotic songs like "The Battle Hymn of the Republic," "The Star-Spangled Banner," and an English-language version of "La Marseillaise," and the men were encouraged to sing and chant on the march. Officers delivered lectures to them intended to fire their blood and read from patriotic literature supplied by the War Department, for instance, Elbert Hubbard's poem "A Message to Garcia."

American training reflected General Pershing's desire to imbue the AEF with aggressive spirit. Fearing that the passivity and defeatism that trench warfare had bred among the French would infect the AEF, Pershing wanted to prepare his soldiers to fight in the open and drive their enemies from the battlefield. The army encouraged eagerness for slaughter by building hatred, gradually indoctrinating the men with a sense that German troops were barbarous and inhumane, hardening their feelings against the people they would face in battle. A War Department training lecture declared:

> We know that in certain Belgian towns young girls were dragged out of their homes into the streets and publicly violated by . . . German beasts. We know that Belgian children clinging to their parents had their hands cut off and their parents murdered before their eyes. . . . We know that the German soldier commonly cut off the breasts of the woman he or someone else had violated and murdered. He wanted them as souvenirs. . . . We know that wounded men have been mutilated and in at least one instance, crucified.

One of the chief techniques for developing fighting spirit was bayonet drill, possibly because General Pershing and some of his officers thought bayonets and rifles could overcome machine guns and artillery, certainly because that was a way of encouraging aggressiveness and self-confidence, making the American soldier see himself, in Pershing's words, "as a bayonet fighter, invincible in battle." A training officer told American recruits:

> When you drive your bayonets into those dummies out there, think of them as representing the enemy. Think that he began the practice in this war of running bayonets through wounded, gasping-on-the-ground and defenseless prisoners. . . . They will crucify some of your men like they crucified the Canadians. So abandon all ideas of fighting them in a sportsmanlike way. *You've got to hate them.*

These methods began to transform young Americans. Lieutenant Bernet wrote from training camp that 90 percent of the men here are from good families and college men, yet in skirmish formation with bayonet fixed and charging a simulated enemy, these fellows are all like madmen."

Yet, some resisted efforts to make them into fighting machines, particularly the conscientious objectors, who presented the government with a numerically small but gnawing problem. The draft law permitted men to avoid combat if they belonged to a well-recognized religious group whose principles forbade them to take part in war. It made no provision for German-Americans who did not wish to fire at relatives and former countrymen, for antiwar socialists, and for others who, for political or philosophical but not religious reasons, opposed killing in battle or participation in war. Eventually, though, the War Department broadened the definition to include men who held personal scruples against war. Draft boards certified more than 20,000 conscientious objectors for induction. Of these, nearly 16,000 decided to take up arms, while 2,599 accepted noncombatant service, were furloughed to work on farms, or went to France as part of a Quaker reconstruction organization. One thousand three hundred ninety declined to take part in any kind of military activity.

To find out what made COs act the way they did, the army sent psychiatrists and psychologists to study them. Its experts reported that one group of particularly obstinate religious and political objectors, although highly intelligent, exhibited a disorder characterized by marked or excessively egocentric personalities. Four hundred thirty-one of this group were egocentric, they said; fifty-eight, inadequate; and five, emotionally unstable. The secretary of war made his own informal survey, concluding that only two of several conscientious objectors he spoke with were "quite normal mentally."

In October 1917, Baker ordered post commanders to segregate COs and treat them with "kindly consideration." This did not prevent soldiers from beating a few of them, jabbing some with bayonets, or in other ways trying to force them to violate their principles. Lieutenant Scott Fitzgerald, for instance, leveled his pistol at one objector and forced him to drill. Eventually, the War Department authorized court-martial for sullen, defiant, and insincere COs, and by war's end, more than 500 had been tried—17 were sentenced to death; 142, to life imprisonment; 299, to between ten and ninety-nine years; and others, to shorter terms. No one was actually executed, although some suffered badly, and a few died in prison. After the Armistice, most of the objectors were freed.

A great majority of those classified as conscientious objectors ultimately agreed to fight. Many, no doubt, altered or violated their beliefs under pressure. But some responded to sympathetic persuasion, including Sergeant Alvin C. York of the Eighty-second Division who, after hours of discussion with his battalion commander about what Christ's teachings meant and following days of contemplation and prayer, concluded that God wanted American soldiers to go to France as peacemakers. York eventually went on to become a legendary figure on the battlefield, killing and capturing a large number of Germans.

Despite these efforts to motivate the troops, expert advisors to the army doubted, as late as the spring of 1918, that the AEF was mentally ready for combat. Psychologists, military officials, and propaganda specialists whom the government convened to study military morale noted widespread ignorance of and indifference toward war aims, harsh and unpopular officers, and chaplains who seemed to favor peace at any price. Members of the U.S. Marine Corps and many of the National Guardsmen and regulars had plenty of enthusiasm for the war, and so did reserve officers, noncommissioned officers, and enlisted men who had been training before 1917 in college military units or at special civilian training camps. But since even regular units were filled with raw recruits and since draftees were joining militia divisions to bring them up to strength, these concerns applied to almost the whole AEF.

In March, one of the motivation experts, Colonel Edward L. Munson of the Army Medical Corps, wrote that many conscripts had no clear idea why they were fighting. A large number were illiterates from a "restricted environment" and of "a lower order of intelligence." Many were foreign born, unfamiliar not only with the ideals of the United States but even with its language. The army's efficiency depended, Munson said, not just on money, munitions, or numbers of men but also on the willingness of American soldiers to fight and if need be die for ideas and ideals. He urged the government to begin a program to develop "the 'fighting edge'—the will to win."

Army Intelligence, Colonel Munson, the surgeon general, and army psychological consultants wanted to establish a program similar to one the German army employed, that placed special morale officers in combat units. They also hoped to coordinate morale-building efforts of the armed forces and civilian agencies, such as the Committee on Public Information. The army general staff resisted these proposals for several months, preferring to leave the responsibility for motivating the troops to division commanders, but by August 1918, a new centralized and more systematic program was underway, using camp morale officers to build esprit de corps among recruits.

When the troops reached Europe, the army continued its efforts to strengthen their motivation. By urging soldiers and their families to write one another and by taking care that troops observed Mother's Day with considerable ceremony, it reinforced the soldiers' sense that they were fighting for home and loved ones. Commanders lined up their men on parade grounds and with flags flying and military bands playing awarded them medals and praised them for fighting well. Colonel George H. Shelton of the Twenty-sixth Division told his soldiers how they were better man for man than the Germans, who were basically cowards and bullies and would try every trick to kill Americans and make them suffer. He warned them not to take prisoners but to kill every German they saw. "Get every one of them you can," he said. "Go into a fight cool of head, but with hatred in your hearts and venom in your bayonets. It may be a horrid thing to say, but every Boche killed brings the end of the war nearer."

Officers sought to instill an offensive spirit by bloodying their men in raids and showing them, sometimes through personal example, how they could survive on the battlefield. Colonel Douglas MacArthur, the chief of staff of the Forty-second Division, left his office in division headquarters and repeatedly exposed himself to enemy fire, an act thought to embolden ordinary soldiers. According to the division chaplain, MacArthur's activities struck other officers as a fine technique for building morale and led Major William J. Donovan to suggest it would be a good thing for the army if some general got shot in the front line.

How did these efforts to build morale affect the outlook of Americans who served in France? What actually made them enlist? What led them to fight and continue fighting?

The Impetus to Serve

Judging by what they wrote to families and friends and from remarks they made in diaries and memoirs, many soldiers joined the AEF for just the reasons the government said they should. Horatio Rogers, a Massachusetts volunteer in the Twenty-sixth Division, had read about German atrocities and developed "visions of a Prussianized world." A private in the same outfit, Joseph Cunningham, was not entirely sure why he enlisted, but German atrocities seem to have stirred him and made his blood boil. Colonel Ogden, the Forty-second Division judge advocate, echoed the president's words. He wrote his family, "We were meant to come in this war to save Liberty for the world. . . . It is the only time in the history of the world that a great nation went to war that did not have to, without any hope or desire for material gain. . . ." Lieutenant Philip Shoemaker wrote from a troop ship sailing to France that he had been glad to leave his home to fight alongside the Allies "for freedom and justice to all." Private Charles Probes, who had left Cornell in his senior year and joined a medical unit, recalled how, on his voyage across the Atlantic, he had written his mother that "we are engaged in a crusade." Several other veterans declared afterward that they had enlisted "to save the world for democracy" or because it was a "war to end wars."

Some remembered particular messages that persuaded them. Private Frank McKeown recalled Uncle Sam pointing his finger at him from a picture on every telephone pole and saying, "I want you." Private First Class William L. Roper of the Thirty-fifth Division, idealistic and patriotic to begin with, had been "deeply moved" by a motion picture portraying the "Huns" brutally treating a Red Cross nurse played by Dorothy Gish. Reports of German atrocities made a deep impression on Lloyd Foster of the Thirty-second Division, who wrote his parents that he hoped the American people would never have to submit to the treatment that Belgium endured. He could now see more clearly, he said, why a war that would cost so much money and so many lives had to come. "We are fighting," he explained, "for better womenhood and manhood and true freedom." Private James R. Bacigalupi remembered that he had been induced to enlist by books like *"Over the Top,"* by Arthur G. Empey (an American who had fought with the British and returned as a recruiter to the United States). Joseph Cunningham recalled seeing the war posters and exciting movies of troops on parade looking "fine all dressed up."

Among those who reported why they joined, love of country seems to have been the single most important reason. A very large number of respondents to the World War I Survey said that a sense of duty; a desire to serve; or "patriotism," "patriotic fervor," or "patriotic fever" had led them to enlist. Yet many Americans professed different or additional reasons for putting on the uniform—particularly, a craving for adventure.

Sergeant Langer remembered almost no serious discussion of American policy or of larger war issues. He doubted that he or others he knew had volunteered to smash the kaiser, to pay back the Germans for their U-boat campaign, or to protect Western democracy in the interest of the United States. "We men," he wrote, "most of us young, were simply fascinated by the prospect of adventure and heroism. Most of us, I think, had the feeling that life, if we survived, would run in the familiar, routine channel. Here was our one great chance for excitement and risk. We could not afford to pass it up." Thurmond Baccus, a self-described country boy, remembered how most young men from his part of Georgia had felt there would be no more wars and that the day of adventure was ending. "Some of us had resigned ourselves to the fact that we would become farmers or work in a mill." Elizabeth Lewis Knight of the Army Nurse Corps was "just crazy to go" to the front with an operating team. "Don't worry about me," she wrote her mother, "for this is an experience of a lifetime. There are many nurses [who] would give anything if they could be here."

Men like Langer and Baccus and women like Knight were willing to risk their lives to see what they called "the big circus" or "the biggest show on earth," or "the game of war." Lieutenant Shoemaker wrote his mother:

> This country is all new to us and there is so much of interest to us. A trip like the one I am having now would cost so much that it would probably be a number of years before I could take it. [T]hen there would not be one tenth as much experience. I am perfectly willing to take a chance, probably an awful chance, to get this experience. . . . If I am fortunate enough to come home alive after the war, I will be able to handle any kind of job, a better man than any one that has not been in the war. If I don't come back, I will of first had a chance to see the country, have a good time and be an AMERI-CAN soldier in the World War.

Of course, many of these soldiers, especially the younger ones, could not imagine that they might die.

Lack of opportunity or disappointment in civilian life impelled some men to enlist, particularly those who entered the Regular Army before the war. Sergeant Ernest G. Gray of the Seventh Division said he joined for adventure but also because he had been laid off from work. Lionel Harmison signed up for the Twenty-third Infantry Regiment at age fifteen to help his mother, a Civil War–soldier's widow. When Elmer J. Stovall, who disliked army life intensely, was asked why he had joined the Fifth Cavalry as a sixteen year old, he said it was because of lack of education and working skills, scarce employment, and low pay. A sergeant in one of the National Guard divisions, Philip J. Fitzsimons, recalled that he had signed up in 1914 because it was cold and the military had a warm building. Harold Pierce decided at seventeen that since he had failed at school and had no other trade, he might as well be a soldier. Warren Donaldson said he had joined the service in 1915 because he wanted to be a patriot and because, since he had no father or mother, "it did not matter if I survived." Theodore T. Juengel entered the army because "I lost my girl, and thought this was a good excuse to get killed."

Large numbers of men enlisted in the armed forces rather than be drafted. When the provost marshal general announced that after mid-December 1917 his office would stop allowing men in the most eligible group to enlist, nearly 142,000 men volunteered in the first two weeks of that month. Some of those who decided to enlist rather than wait for conscription would probably have accepted induction willingly. Still, anxiety about being drafted hastened several men to the recruiting station—like Lieutenant Bernet, who said he signed up "not because I was intensely patriotic but simply because . . . waiting for the publication of those draft lists would have set me crazy."

Some immigrants and second-generation Americans joined the army to repay an obligation they felt to America. Service in the armed forces could also enable immigrants to become citizens more quickly. Knud J. Olsen, a Danish-American in the Eighty-second Infantry Division observed, "I earned my right to citizenship by being willing to die for my country." As we have seen, some people of African ancestry and American birth enlisted to be recognized as having the rights of other American citizens.

American schools encouraged their male pupils to enlist when they reached military age, teaching national allegiance and, in history classes, the glory of war. Private Harland King had left his job as a toolmaker, he said, and joined the regular army because he loved American history. William S. Parker, another regular, stated that "from the reading of history I had always wanted to serve in a war."

Then there were the men with war in the blood, the street fighters from the slums of New York, the descendants of warring Celts, and the "bearded ruffians" and bar-room brawlers from the corn fields and coal mines of southern Illinois and the violent moonshine country of the Kentucky mountains, the men of the Thirtieth Division, whom one of their number described as "typical red blooded, normal country boys, who spent most of their spare time scrapping among themselves & harassing the M.P.'s, . . . brave, ingenious & loyal scrapping young men from the South, 99% Anglo-Saxon." And finally, there were all those young Americans, including some of those just mentioned, who, half a century after the last big war and despite the movies and the war stories, were totally innocent of what the battleground was like.

While Americans were moved as individuals to enlist in the armed forces, they often joined as members of groups and were influenced by their connections to others in their hometowns or neighborhoods or schools or colleges or families. Private Walter Christensen recalled how his classmates had gone on a picnic and talked about the war and how all the boys had decided to go into some branch of the armed forces. The members of William Briggs's law-school class were so eager to join that they took a train to a nearby city and signed up the day war was declared. On April 17, Duncan Kemerer of Export, Pennsylvania, his brother, and six friends joined the Pennsylvania National Guard. Eleven days after that, fifteen-year-old Homer Wilson and a small group of boys about the same age went to the National Guard office in their Texas hometown and enlisted. Others remembered entering the army because "everyone was doing it," because all their friends had joined, because they wanted to be with someone who had gone in, or because of what John D. Clark, an Amherst student, called "war hysteria."

The inclination to serve was overwhelming at places like Amherst, Princeton, Yale, Harvard, and other colleges and universities across the country where young men from privileged families had been preparing for war for years. Some college students had gone to military prep schools or belonged to old martial volunteer organizations. Many of them were shocked by the ruthlessness of the German armed forces—their invasion of Belgium and the atrocities they were said to have committed there, their sinking of neutral ships and killing of civilians in bombing raids and U-boat attacks, and the chemical warfare that German troops employed— and looked for a way to take sides.

Hundreds of college students had volunteered to drive ambulances for the French army or had taken part in summer camps that the army chief of staff General Leonard Wood inaugurated in the summer of 1913. Others had trained under the auspices of the Military Training Camps Association, founded by Wall Street lawyers and investors and other members of the Anglo-Saxon Protestant elite after the sinking of the *Lusitania*. The MTCA had arranged with General Wood to establish camps at Plattsburg, New York; Fort Sheridan, Illinois; the Presidio in San Francisco; and American Lake in Washington. In the summer of 1915, thirteen hundred trainees assembled at Plattsburg, most of them from East Coast colleges and clubs. The group included the mayor and police commissioner of New York, the Harvard football coach, the Episcopal bishop of Rhode Island, three of Theodore Roosevelt's sons and one of his cousins, a partner in the House of Morgan, and war correspondent Richard Harding Davis.

Aside from their feelings about the German Empire, the college-educated elite had several reasons to promote and take part in military training. Many of them felt an ingrained obligation to fight for their country out of class and family tradition and personal code. "To be young and not in uniform," wrote George A. Vaughn, an army flyer who had enlisted with some twenty-five other Princeton students, "in those days was a disgrace."

Some members of this elite believed that if men of their own kind shared tents with young workers, they could soften class differences that threatened to divide the nation. Several hoped military training would homogenize immigrants into a purer American type or saw it as a way of imposing order on an undisciplined society. While many were captivated by romantic feelings about war, several felt they needed military service to harden and discipline themselves and people like them whom wealth was making soft, and some hoped it would toughen their countrymen. "A war is good for the country," Lieutenant Bernet remarked. Bernet, a fraternity man from Kansas State University, believed "a scourging once in every so often is a benefit rather than a hindrance" and that the United States would emerge from the war as a country whose people would "for the first time appreciate the flag, and . . . know how to cooperate as a nation in commerce and industry." Several of the pro-preparedness elite wanted to train themselves to become leaders of an army that would fight alongside allies—Great Britain and France—to which they were tied by tradition and blood.

Blood and group tradition or a sense of obligation influenced many Americans to join the armed forces. Sergeant Major Vernon Mossman explained that he had volunteered for the First Division because he came from a family of patriots. Corporal Edward Schafer wished to pay back the United States for the opportunities it had given his German immigrant parents, while Harold Cline enlisted in the Student Army Training Corps so his family could do more for the war effort. It was not just for himself that he had joined the National Guard, said Corporal Merwin Palmer; it was to carry on his family's duty to the nation.

A large number of the recruits of 1917–1918 had ancestors who had fought in earlier wars, particularly the American Civil War; and while military heritage did not guarantee that men would enlist, it influenced a good number do to so. Corporal Palmer explained that he had joined because his paternal grandfather had lost his left arm in the Civil War and his maternal grandfather had been the first man from New London, Connecticut, to enlist in that conflict, and he wanted to follow their path. When Clarence Mahan learned that the United States had gone to war, he kept remembering that his paternal grandfather had fought for the Union and wondered what he should do. Private Rivers E. Banks enlisted in the Seventh Engineers to be as brave as his grandfather, a veteran of the Grand Army of the Republic. George O'Brien, whose ancestors had fought as captains at Waterloo and in the Crimean War, joined the American Expeditionary Force to uphold his family pride.

Social and family ties led away from war as well as toward it. Mennonites and other religious conscientious objectors followed centuries of antiwar tradition. When Corporal Schafer visited his favorite tavern after deciding to join the army, other German-Americans he met there wanted to know "if I really was going to fight the Fatherland." His neighbors found it hard to understand that people whose families had come to America to escape conscription in Germany were going to war for the United States. As another German-American, Rudolph Forderhase, drove away from his family's farm to report for duty, he looked back and saw his father with bowed head, sitting on the farmhouse steps, "utterly dejected." Claude Hopkins's brother had joined the Forty-second Division and returned home on leave, wounded, gassed, and dying from pneumonia. He told Claude not to enlist. Even though Hopkins had been trained in the Reserve Officers Training Corps, he did not volunteer but waited until the draft reached him.

Some families who did not want their sons to fight or had divided feelings about it tried not to let on how they felt. Horatio Rogers observed his parents' silent anguish while he was trying to make up his own mind. Their suffering, he said, offered a reason not to enlist that "appealed to my cowardice," but he still felt uneasy about staying home and joined the National Guard. A Pennsylvania farmer's wife was much more outspoken than Rogers's parents. She told her husband's draft board, "My George ain't for sale or rent to no one. If he goes, I got to go too." Assurance that she would receive his allotment failed to pacify her. "I don't want your money," she said. "I jist wants George." When George failed his physical examination, she turned to the chairman of the draft board and stuck out her tongue.

Other families, however, urged their young men to enlist, even as they regretted their going. A woman from Jefferson County, Arkansas, wrote her husband's draft board, "I will not hold a man needed so badly by his country if I can possible help it. I love my husband very much and hope he will kill many Huns, help win victory and come safely back to me." When hysteria was rising against German-Americans like William Koester's family, his father insisted that he join the army. In some families, boys competed with their brothers to be first in the service or enlisted simply because it seemed the normal thing to do after their relatives had gone. Private Lyle McDonald entered the army at age seventeen to help somehow his brother who had been reported missing in action.

The Impetus to Fight

Attitudes that impelled young men to enter the armed forces were part of what kept them fighting. Thousands entered combat touched by joyous, romantic feelings. "Our boys went [to] the battlefield last night singing . . . ," a regular-army soldier told his sister during the Aisne-Marne campaign. "[T]hey are surely [a] game and happy bunch." Raymond B. Fosdick wrote his family from the front in early June 1918 that he had seen an American division going into action with red poppies on their helmets. "They swept by like plumed knights, cheering and singing. . . ."

Traditional incentives spurred them into battle and sustained their will. Private Sweningsen told his mother that he wanted a medal for bravery. Martial music thrilled them, even after grueling combat. Captain Philip McIntyre recorded how the men of his division threw out their chests as a band played "The Star-Spangled Banner" and "every man's eye had a new glint in it," while McIntyre himself "just tingled from toes to head." Alpha C. Miller of the Sixth Marines thrilled to march to "The Stars and Stripes Forever," and Corporal Harold Pierce, back in a rest area after days of draining combat, found himself stirred and renewed by the marches of a military band.

Members of the AEF felt that the country's idealistic and altruistic aims had inspired them in battle. They remarked that they felt God was on America's side, that they were fighting in a just cause, and that they were bringing democracy to the world or liberating France. Some said they fought to cleanse the world of Prussian militarism and autocracy, to prevent a tidal wave of German barbarism from sweeping over civilization, or to repay the "Hun" for atrocities against the women of Belgium. In a letter given to his mother after he died in a crash, Osric L. Watkins of the 94th Aero Squadron wrote, "With Liberty and true Christianity at stake you would never think of shrinking from the sacrifice." It should be noted, though, that at least in parts of the American Expeditionary Force it was considered bad form to seem too idealistic and that several men, who said they were fighting for high-minded purposes, made such comments before their first battle or long after hostilities ended. More personal, more concrete considerations probably motivated most American combat troops.

For some, a driving objective was to acquire mementos of the war. Men risked their lives to retrieve enemy helmets, weapons, and insignia. Lieutenant Quincy Sharpe Mills crawled into no-man's-land to pick wild roses from a bush and sent a handful to his mother. Our people, wrote Corporal Edmund Grossman, were "crazy for souvenirs."

Some of them, at least in the beginning, feared punishment if they failed to fight or ran away. "[W]e didn't want to get in wrong with our C[ommanding] O[officer]s," Private Aitken wrote, "and get court-martialed for desertion or some such thing." Some considered combat "a supreme test of manhood," as Private Baccus put it. Other soldiers spoke of doing "a man's job" and of the way "real men" performed in war. Many regarded fighting essentially as a job to be finished before they could go home. Responding to the World War I Survey question about what he was thinking during combat, Private Lowery F. Aden said, "Going through Hell mud rain hungry . . . trying to do a job and stay alive." Corporal George O'Brien answered, "It was an adventure and a job to do"; and Lieutenant William S. Parker declared, "Scared to death . . . but the job had to be done." Homer Wilson responded that war was worse than hell, and "the sooner we got it over the better."

A number of men remembered not thinking at all, certainly not reflecting, but acting mechanically, impelled by instinct or the excitement of battle or ingrained routines. Duncan Kemerer had felt like a robot, nerveless until the battle was over. Lieutenant Allen recalled how soldiers in combat lived from moment to moment with an "animal keenness" that utterly absorbed them. This was fortunate, he thought, because they would have found their circumstances intolerable if they had time to brood. Private Wilson remembered fighting just to kill or be killed. To such men, ideals and conscious motivations had little significance at the time.

The excitement of combat was a motivation to fight for some, and even men who suffered greatly found themselves craving the high it produced. Malcolm Aitken, a private in the Second Division whose memories of battle gave him nightmares, wrote his father eight months after the Armistice that though he hoped the war did not start up again, because that would delay his return, he would be only too glad to go to the front if it did resume, "as we are just aching for a little excitement. . . ."

Just exposing troops to fire prompted them to join in battle. At one point, Private First Class Knud Olsen recalled, he had two German soldiers and an officer in his sights. "I was hesitating," he wrote, "because I didn't really know if I wanted to kill someone. Then I heard a bullet whistle by my head." Olsen fired off a shot, and the three men hit the ground and began crawling back to shelter. The incident changed Olsen's attitude toward killing. From then on, he enjoyed shooting at the enemy.

Although, as we have seen, front-line troops had varied feelings about the enemy, some men's hatred of the Germans intensified their desire to fight. Lieutenant Mills wrote his father, "The more I see what the Germans have done over here, the more I long to kill some of them." Mills, who censored the letters of the Iowans who served under him, noted they were writing home that they wanted to punish the "Boches" just for bringing them to France from "God's country." They did not hate the Germans before they came over, he told his parents, but now they displayed in manner rather than words "an increasingly grim determination to 'give Fritz hell.'"

Some doughboys considered Germans subhuman and therefore suitable for slaughter. Arthur E. Yensen, a wagoner with the Fifth Division claimed that American troops had been "brainwashed into thinking all Germans were rats that ought to be killed," while Hervey Allen saw them as ruthless ant-like people who had crushed out every trace of individualism in themselves and fought with the bravery and determination of insects. Jasper Goodson, a wagoner in the Thirty-seventh Division thought of the Germans as "objects to be eliminated." Captain Douglas Campbell of the 94th Aero Squadron remembered feeling that the flyers he battled in dogfights were not people at all but "dangerous birds who would get me if I didn't get them. . . ."

From a place that still smelled of corpses, Leo Culbertson wrote that his heart had become so hardened that seeing dead Germans gave him pleasure, while the sight of dead Americans being buried in their uniforms made his blood run cold and strengthened his "passionate desire to deal out misery to the enemy." Gary Roberts, a soldier in the Forty-second Division, wrote from a hospital: "I got two (2) of the rascals and finished killing a wounded one with my bayonet that might have gotten well had I not finished him. . . . Why I just couldn't kill them dead enough it didn't seem like. Believe me it was some fun as well as exciting. . . . The first one I got was for momma and the other one was for myself for the troubles they have given this old boy since I left Home." An army flyer, Lieutenant Hamilton Coolidge, told how thrilling it had been to strafe German troops the first day of the Argonne offensive:

> How my old heart hammered with excitement as I dove down beside that road, not fifty feet high, and recognized those Boche helmets! In a twinkling I was past them, gained a little height to turn in safety, and came diving down upon them from the rear. I just held both triggers down hard while the fiery bullets flew streaming out of the two guns. Little glimpses was all I could catch before I was by. Another turn and down the line again. I had a vague confused picture of streaming fire, of rearing horses, falling men, running men. . . . I found myself trembling with excitement and overawed at being a cold-blooded murderer, but a sense of keen satisfaction came too. It was only the sort of thing our doughboys have suffered so often.

Conditions on the battleground, particularly the deaths and wounding of other Americans, stirred the desire to kill. Gerald Gilbert was ready to slaughter every German after enemy shells landed on a hospital where he worked. The Germans, he thought, were unprincipled. They did not know right from wrong. They should be exterminated. "To see blood and carnage everywhere," Corporal Clarence L. Mahan wrote, "as men, horses and mules are blown to bits, developed in us a certain savagery and hate that pushed us on toward a terrible enemy with a willingness to see him destroyed." Private Strauss described how one of his officers found a German sniper in a tree in the Argonne forest shooting at American walking wounded. He yelled at the German to come down, "you bastard!" and the German did come down, crying "Mercy, Mercy"; but before he reached the ground, the officer put two revolvers to the sniper's face and blew his head almost off, saying, "I'll give you mercy you bastard!" When the Germans killed a friend of Sergeant Lyle S. Cole, it made him so angry that all he could think of was "getting to the top and getting at those Dutch Bastards."

As Sergeant Merritt D. Cutler of the 107th Infantry Regiment saw more and more Americans go down during an attack on the Hindenburg Line he developed an "incredible desire" to destroy a German pillbox. He worked his way around one and dropped a grenade in the rear opening. "I'm not a violent man," he later remarked. "But I can remember my glee when the grenade exploded. It horrifies me now, but I was in such a state of shock that nothing mattered but the killing of some of those Germans."

These actions, stimulated in part by a desire for revenge, reflect the bonds that had grown between groups of combat soldiers. The men came to love one another and to admire and respect each other's bravery. They had also quickly learned how their lives depended on cooperation and how terrible it could be for a man if his buddies felt he had let them down. Honor, or dread of shame, became a powerful incentive to fight. Corporal Pierce, having survived the carnage in Fismette—the "Valley of Death," he called it—was "in a hurry to leave this misery and death trap" but did not want his comrades to know it. Sergeant Cutler recalled that during the San Quentin Tunnel attack, he concentrated more on moving ahead and not letting his buddies think he was "yellow" than on the Germans who were shooting at him. Private Aitken told how at his first battle he was gripped by an extraordinary fear but took courage

> since the other fellows were not showing yellow and stuck with the bunch. It was that very thing that kept us together. All of us were afraid in a sort of way, in that we didn't know what we were getting into and didn't know what to expect. But in order to keep our personal reputation up, that is the little bit that we had been able to build up in the last few days . . . , we were more afraid to go to the rear than to the Front.

These men found themselves fighting simultaneously for their loved ones at home and for their comrades. Private First Class Walfred Walker told how sad he felt when the time came to leave his buddies, "boys we had learned to like as brothers of a large family."

Bonds of comradeship were especially strong in organizations that had fought together for months or years, among troops who had volunteered and knew that their buddies had volunteered, and in units whose members had come from the same colleges and high schools or the same clubs and fraternal societies or who had grown up together in small towns or neighborhoods. Sergeant Robert Cahal of the Thirty-sixth Division had been acquainted before the war with thirty-five or forty men in his company, boys from small sawmill towns and country areas near his home, a congenial, happy, friendly group whose families knew one another. It is easy to imagine the pride their relatives felt in these young men and the community shame that would have fallen on any of them who were found to have run from battle. Corporal Shafer of the Twenty-sixth Division explained that since everyone in his unit was from the same town and maybe the same school and church, a man could not possibly be at ease with himself if he deserted his comrades.

Comradeship had its terrible side. Captain Philip McIntyre told how preparing battle reports caused him pangs and heartaches, bringing back fresh memories of the deaths and wounding of "my brave boys and officers." When a buddy of Lieutenant Douglas Campbell of the First Pursuit Group died in a landing accident, Campbell was heartbroken. "[Y]ou played a dangerous game," he remarked many years afterward, "by becoming too fond of the other pilots—you paid a heavy price when they caught it." But it had been hard for him not to form attachments because his group was "about as great a bunch of fellows as you could ever find. . . ."

For a number of the soldiers in the American Expeditionary Force, the strongest motivations to fight, even more powerful than comradeship, were ties to kin, especially to parents, and above all to mothers. These men felt they were fighting for their homes and for the reputations of their families and because their families implicitly or explicitly required them to enter the battleground. Ernie Hilton of the Eighty-ninth Division wrote that his folks at home were "really the reason why I am over here." Sergeant Cyril B. Mosher, a deeply sentimental young man from Kentucky, who referred to himself in letters to his mother as her "little boy," told her he thanked God he had such loved ones at home and said, "it is for your safety and happiness that I am here." To his sister, he described how he had slept in a French house with "all those little things which make a home; and I thought how great a blessing it is that this war is being fought outside the houses of our dear ones. It is such sights as these that show the American soldier what he is fighting for."

"What makes the soldier fight?" Norman Summers asked. Sergeant Summers, an infantryman in the Rainbow (Forty-second) Division, was in the hospital, having breathed too much gas while waiting to go into the line near Château-Thierry, and he had time to ruminate. His answer: "It is because our people back home expect it of us." Thoughts of his mother, he said, had put courage into him that he could never have shown if it had not been for her:

> What made me go into the front lines to face the enemy and help drive them back when I had the opportunity of staying in the rear practically out of danger? And after I had helped drive them back for two days and nights and was again told I could go back why didn't I go? I was at no time compelled to

go to the front but yet I went with my company and for just one cause—for my mother's and father's sake. They would have bid me go. They wouldn't want to say after the war is over that their son didn't go into the front lines and now they won't have to say it.

Summers thought about the many times he had wanted to run as fast as he could to the rear when the shells were raining around him, "but the picture of Mother, her firm chin set, and her eyes looking straight through me seemed to say, 'Son, stick it out.' Then my courage would bound up and I would no longer fear the biggest shells the Germans could put over. I am not brave—my mother is—and I believe her prayers have saved me this far. . . . My earnest prayer is that I may be near as possible 'The Boy my Mother thinks I am.'"

Other parents, obviously worried about losing their young men, encouraged them to keep fighting. Horatio Rogers's father, Reverend Lucian Waterman Rogers, who months before had seemed anguished over the thought of his departure, sent a letter that reached him in the front lines: "I wish to heaven that I were in your place. The great call came to you and it came to my father, but I am left behind to nibble around these trees like an old rat. . . . Here's to peace through victory." George H. O'Brien's mother told him that hard as his leaving had been for her, she would not have had him stay at home. When he did come marching home, she wrote,

bring me back the same boy I gave my country,—true, & clean, & brave. You must do this for your father & me & Betty and Nora;—& most of all, for the daughter you will give me one of these days!. . . Live for her or if God wills, die for her;—but do either with courage. . . .

O'Brien assured her that he would give a good account of himself "so you won't have to be ashamed of me."

The Disillusionment Question

Combat affected the values of soldiers in ways quite complicated and often misunderstood. During the 1920s and 1930s, John Dos Passos, Laurence Stallings, William Faulkner, and other American writers depicted the First World War as a grim, brutalizing conflict in which ideals did not survive very long. Reading their works, the so-called literature of disillusionment, one might believe that the Americans who did the fighting lost all the romantic visions they brought to France.

Yet for many of the eager, cheerful, optimistic young men who set out for the Old World in 1917 and 1918, that clearly was not so. Fifteen years after the Armistice, veteran Ramon Guthrie wrote that he had enjoyed the war, the close association with comrades in a common purpose, the chance to put forth "intensive and disinterested effort in a cause greater than one's own personal concerns," the freedom from economic cares, the equality of rich and poor, and the adventure. William Roper, who had signed up because he was adventurous and patriotic and idealistic and wanted to make the world safe for democracy, remarked long afterward that he had relished the war, the sight-seeing, the excitement, and the "learning experience," everything but front-line combat. It had been, he said, "My Great Adventure." Other veterans recalled that the Great War had taught them how wonderful America was, that it was the greatest country in the world and that its people were the finest anywhere. Sergeant Sol Cohen of the Thirty-third Division remembered feeling while the war was going on that he had played an important part in the liberation of France and "exulted in it all." Cohen had been and continued to be "a willing disciple of Woodrow Wilson."

Other veterans remembered the war as painful and sweet. Walfred Walker spoke for them when he returned to Dickinson, Texas, in June 1919 after fighting in a "living hell." He was glad to get home, he said, but "found it mighty lonesome to part with friends with whom I had soldiered for two years. All in all . . . it was a great war—with happiness intermingled with sadness."

The war had no obvious impact on the feelings of some men. Heywood Broun, a journalist who accompanied Pershing's army overseas, remarked in a highly critical review of Dos Passos's *Three Soldiers* that although war was brutalizing, "the most terrific adventure and ordeal the world has ever known lifted its hordes in the air and when it set them down again vast numbers had not been changed spiritually by so much as a single hair. . . ." Considering how many servicemen never approached the front lines, Broun's claim seems plausible. Yet disillusionment, a changed attitude about war or the nation or about the hopes with which the war had begun did occur, sometimes long after the war, sometimes on the battleground itself.

Certain men appeared particularly susceptible to disillusionment, so much so that one has to wonder whether they had ever accepted the official ideals. Sergeant Forderhase—a draftee who titled his diary "We Made the World Safe!!!!?????"—said after the war that the waste of life had been "tragic." Yet he made it clear that he had gone to war in the first place with little enthusiasm. Henry Fritscher, who afterwards described newspaper coverage of the war as mostly propaganda and called histories and articles about the conflict "fiction glorifying war," remarked that as a child he had hoped for war so he could join the army. But the military, he said, "made a super Patriot into a rebel." This suggested that he had been an eager fighter who lost his ideals. However, Fritscher may never have been a really enthusiastic warrior since he also remarked that the army had to "brainwash" him to make him a soldier.

Yet certain veterans did lose their illusions—like a former private who began the war as a patriotic fifteen year old, eager for combat and proud to serve, and ended it sterile from an attack of mumps and "without much hope"; and like Corporal Clarence Mahan, who after "crusading for freedom" finished the war with memories of the wounded, the dead and dying, and the hate he had felt in battle. Willard M. Tyner, who entered the army as a draftee, spoke for many of his generation when he recalled that "we were all very patriotic" but came to feel "we were suckers to get into that war. . . ."

Accounts of the war that appeared long afterward may reflect disillusion that took years to set in. Harold Pierce's "Diary of a Doughboy," telling how the author's spirits were desolated, reinvigorated, and crushed once again by the events of the war, is one example; another is *Wine, Women and War: A Diary of Disillusionment,* by Howard V. O'Brien. In this book O'Brien wrote, "It's funny how notions change. Back home we drooled about 'democracy' and 'glory.'" But now the people the newspapers said were "eager for the Front" were "about as eager for the smallpox." These remarks may have represented what O'Brien saw in France in 1918. They surely expressed his own feelings about the war in 1926, the year his book was published.

Sometimes, we can see feelings about war changing in the midst or the immediate aftermath of combat. "It is all like a great cinema," wrote Lieutenant John D. Clark, a field artillery officer at the beginning of a battle, "constantly changing, constantly moving." Later the same day, he added: "It is sad to see the wounded in the hospital, but not nearly so gruesome as on the field when both dead and wounded are strewn about smeared with blood. I am just beginning to realize what war really is." Horatio Rogers remembered how a sergeant in his battery, "the most reliable, hard-working sane human I ever saw, and a giant of a man, admitted he had 'lost interest in the war,'" and that he himself had begun at that time to write letters home "full of . . . artificial optimism." Lieutenant Shoemaker, who once had talked of fighting for freedom and justice, wrote his mother while he was recovering from wounds, "I am well fed up on the killing game, and most anxious to start making things grow and animals reproduce." Private Stenback, who had joined the marines out of patriotism and had wanted to go to France so badly that he preferred to be a buck private over there than a sergeant in the United States, told his father a month after the Armistice that he had seen his share of the war "and have had enough of it." Lieutenant Lawrence, who had started the war adventurous and imbued with the justness of the cause, recorded in his diary for November 10, 1918, "We were not enthusiastic about another battle, for the last had stripped war of its glory. . . ."

These altered feelings may have been for some a temporary response to battle stress. But for certain veterans, disillusionment lasted. In 1939, a few weeks after Hitler's forces attacked Poland, Lawrence wrote: "There cannot be any enthusiasm for another war—there are too many maimed men still living, too many war-made widows, and too many mothers who still live whose sons went to war and did not come back."

The attitudes of veterans were affected by what happened to them after they returned. Several respondents to the World War I Survey described themselves as forgotten men who had received no rewards from the government they had served, certainly far less than veterans of later conflicts. Malcolm Helm, a U.S. Military Academy graduate who had fought as a captain in the Second Division, wrote a half century later "that a person who entertains soldiers or portrays a military hero is honored or decorated higher than men who went through machine-gun fire, or artillery barrages and faced flame-throwers or poison gas." And George O'Brien, whose mother had admonished him to come back clean and brave, recounted that he had received no benefits from his government, even though he had spent two and a half months hospitalized and was crippled by gas. "Americans will double-cross you any time they get a chance," he wrote. "Americans are looking for the Almighty $$$$$$. Many are willing to take its advantages but want others to fight for them." O'Brien had learned from the war, "to trust no one."

We can now make some generalizations about what motivated the men of the American Expeditionary Force and about the effectiveness of government efforts to encourage Americans to fight. Those efforts frequently paid off. The official war aims and the atrocity tales disseminated by the government were mirrored in accounts American soldiers gave of why they went to France and what they were fighting for. Through military training, the government evoked and channeled hostile feelings toward the enemy. It coerced or persuaded certain conscientious objectors to abandon their qualms about killing or at least about military service. By placing troops in a combat situation where they had to kill or be killed and putting in their midst officers and noncoms to direct, urge, and inspire them, the American government prompted them to advance into fire.

For American fighting men, the influence of groups was very strong. They entered military service and fought because they thought it natural to do what their peers were doing or their relatives had done, because an ancestral tradition had to be upheld, or because the honor of their family and its place in the community were linked to the way they behaved. Many soldiers' willingness to fight was sustained by the small group of buddies with whom their lives were intertwined. The government reinforced their bonds to families and friends as a way of encouraging them to kill and conquer and endure the pain of battle.

Yet doughboys were not simply the instruments of higher authority. They brought to military service attitudes and desires that helped others transform them into an army. Some were young, with a boy's ideas of war shaped by adventure stories and by fables of military history taught in school. Many longed for adventure, wished to test themselves, or saw battle as a ritual to turn them into men. Patriotism appears to have inspired a great many of them, though the meaning of patriotism varied considerably. For some, battle itself aroused an impulse to fight. Some felt they had to fight because there was no alternative.

The type of motivation and its strength and persistence varied greatly among American troops, from soldier to soldier and from unit to unit and from time to time, sometimes lasting through weeks of combat, sometimes breaking down. It is to those whose fighting spirit dissolved under the stress of battle we now turn and to the government's efforts to heal their minds and send them back to the battleground.

12 ||| The Treatment of "Shell-shock" Cases in the AEF: A Microcosm of the War Welfare State

In the first weeks of the Great War, an apparently new kind of casualty appeared on the battlefield, a soldier with staring eyes or a terrified look accompanied by violent tremors and cold, sometimes blue, extremities. Some of the victims had become deaf; others, dumb; some were blind or paralyzed or had combinations of several of these symptoms. At first their syndrome was attributed to shelling. People called it "shell shock" because they believed concussion of high explosives had shocked the patient's brain.

When the cases were looked at more carefully, this explanation did not hold up. The symptoms were far more likely to appear in men with no wounds at all than in those with head wounds and damage to the central nervous system. Prisoners of war developed shell shock far less frequently than their captors, even when both groups were subjected to the same bombardment. Military psychiatrists concluded that the origins of shell shock were often psychological and that its symptoms, which resembled some they had seen in civilian practice, had been touched off by incidents of the war. The U.S. Army Medical Department called the cluster of mental illnesses that had not arisen from physiological damage "war neurosis" or psychosis. Ordinary soldiers continued to use the term "shell shock," however, and so shall we, recognizing that this was the common parlance, not a precise medical term.

In June 1917 Dr. Thomas Salmon, one of the nation's chief mental-health experts, sailed for England to learn more about shell shock and about the way the British and the French were treating it. A former Public Health Service psychiatrist, Dr. Salmon was medical director of the National Committee for Mental Hygiene, an organization dedicated to improving the mental health of the American people and the care of the mentally ill.° Committee members felt that psychiatry had an important part to play in what Salmon called "the great movements for social better-

---

°The National Committee for Mental Hygiene had been founded in 1909.

ment"; and like other professional groups and other social reformers, they volunteered their knowledge to the war effort. As a member of a group of psychiatrists and neurologists called to Washington by the surgeon general, Salmon helped create a program for preventing and caring for psychoneurotic casualties in the army. Later he became the chief psychiatric consultant to the American Expeditionary Force.

Salmon's report from England contained ominous facts and observations. He noted that mental and functional nervous disorders were responsible for at least one out of every seven disability discharges from the British army and estimated that the annual rate of admission to hospitals for mental illness among British troops returned from overseas was four per thousand, quadruple the rate for adult civilian males. These diseases, he remarked, endangered morale and discipline in a special way and required attention for purely military reasons. He meant by this that what the soldiers called "shell shock" could be used by troops, intentionally or otherwise, to escape from combat, undermining their country's fighting power.

Many soldiers in the Allied and American armies believed that shell-shock victims—men who ran around crazed by shelling, cursing and raving, jabbering and dancing about, struck dumb or screaming or crying, shaking uncontrollably or paralyzed, going into convulsions and frothing at the mouth, or staring blankly and hollow-eyed as in a trance—were as ill as those with bleeding wounds. They also noticed that anyone was prey to shell shock, that the strongest-looking men could be the first to crack and that a man could perform heroically one day and fall apart the next. "All of them are not cowards," wrote Sergeant Summers, of the Forty-second Division. "I have seen some men who are recommended for the Croix de Guerre on other fronts go to pieces later and be sent to the rear." It happened to a man in Lieutenant Lawrence's unit, "one of our best sergeants," who broke after four days of shelling and cringed in the commander's dugout, "his nervous system shattered . . . whimpering like a baby; he was a fine young fellow and a brave soldier, but there was a limit to human endurance." Since the shame that attached to leaving one's comrades for the rear did not apply or did not apply as fully to a man who exhibited these symptoms, shell shock corroded the bonds that held troops to one another on the battle line.

A number of the signs of this disease resembled normal reactions to combat. The fatalism soldiers acquired after days of heavy fighting appeared like and sometimes turned into depression. Willingness to expose oneself to fire might be hard to discriminate from attempted suicide. Battle dreams so terrifying that men stayed up night after night to avoid them not only plagued shell-shock casualties but occurred regularly in front-line troops. Indeed, one distinction between the ordinary combat soldier and the shell-shock victim was that the former's symptoms abated without treatment.

While trained physicians could detect a pattern of mental disability among victims of neuropsychiatric disease, troops often misdiagnosed particular symptoms they observed in themselves and thought they had it when they did not. This led to shell-shock epidemics as men with real or imagined symptoms began to escape the battlefield. If soldiers who left did not return quickly, the disease might break out among those who remained. But if a man with the symptoms stayed at the front untreated, he could set off an epidemic as others consciously or unconsciously mimicked him. Besides, it was futile or worse to try to force stricken soldiers to remain in the lines just because they might be malingering. The illness was often truly incapacitating, and to keep someone in combat with a serious psychoneurosis not only invited his death but endangered his comrades and interfered with their ability to fight.

American military psychiatrists soon found themselves facing the same delicate problems as their French and English colleagues—how to prevent psychological incapacitation, how to distinguish genuine cases from simulated ones, and how to treat mental illness among soldiers without encouraging its spread. At the suggestion of Dr. Salmon and other experts, the army tried to keep the "mentally unfit" out of the service; but it was not possible to exclude all of them before induction or during training. More than eleven thousand psychoneurosis cases were identified in the camps, and several cases, identified or not, made it to France, where as late as mid-July 1918 General Pershing was complaining about the "prevalence of mental disorders in replacement troops."

Salmon anticipated that, regardless of screening, large numbers of American soldiers would become mentally disturbed and recommended that the AEF prepare a system for managing them, including allotting one hospital bed for each thousand combat troops. The army responded by assigning division psychiatrists and by setting up special neurological hospitals near the combat zones, a special thousand-bed Base Hospital no. 117 in France for war neurosis cases, and other centers overseas and in the United States for treating neurological and psychiatric casualties. With the experience of the French and British in mind, the Army Medical Department tried to have psychological and neurological injury cases diagnosed and treated as speedily and as close to the front lines as possible.

Official enumerations of these cases were not very precise. In 1917 the American Medico-Psychological Association and the National Committee for Mental Hygiene agreed on systems for classifying mental and neurological illnesses and injuries, and the Army Medical Department adopted them for its own use. But some of the categories into which doctors placed their patients were fuzzy, like "Shell Shock (Cause not given)" and "Paralysis without specified cause." Patients with similar diseases may have been classified differently. Qualities of similar intensity that may have labeled one man as a constitutional psychopath—instability, undependability, changeableness, emotional unbalance, and asocial behavior—might have been diagnosed in another as psychoneurosis, depending on the physician and the circumstances.

While it is impossible to say exactly how many soldiers were afflicted with these mental diseases, some notion of the general problem of mental and neurological illnesses and defects in the American army is suggested by the following figures in the Army Medical Department history: 11,000 psychoneurosis cases identified in training camps, 8,319 cases returned from overseas, 97,497 admissions to army hospitals, 41,976 disability discharges for neuropsychiatric diseases in the AEF.

Case histories illustrate what army psychiatrists described as war neuroses and what ordinary soldiers called shell shock. Private L. J. A., a twenty-year-old former mill worker from Massachusetts arrived in France with the Thirtieth Division in April 1918 and went to Château-Thierry early in June. On July 14, he came under shell fire, which made him nervous, though he was able to carry on for eleven days. Then, during a heavy barrage, he lost consciousness while digging a shelter to protect himself. When he awoke about five hours later, he was in a hospital, dizzy and shaky with pains in his eyes. His diagnosis was neurasthenia, a neurosis characterized by intense fatigue. He was one of the nearly nine thousand soldiers admitted for treatment of this disease.

Private D. E. R., a motorman in Lynn, Massachusetts, enlisted in May 1917 and went to the front lines early the following year with the 101st Ammunition Train of the Twenty-sixth Division. During the second week in March, someone hit him on the head accidentally with a rifle, but he recovered easily. On the 23rd, he was in Soissons during a German bombardment. A shell struck about fifty yards away, and while it did not appear to hurt him, he became terrified and began to shake from head to foot. A day later, he started twitching all over, several times a minute. He lost his appetite and slept poorly, awaking with a start from nightmares in which he imagined himself in a cellar with shells blasting above. The left side of his head, where the rifle had hit him earlier, started to ache, and he jumped at the slightest noise. Two months later, his condition was essentially the same. He was diagnosed as suffering from psychasthenia, a neurosis in which one cannot make a choice for fear of the consequences of choosing. Army physicians noted more than eight hundred cases of it.

Sergeant O. C., a draftee in the 362nd Infantry, never made it to the front. He came down with influenza and was admitted in April 1918 to a hospital, where for three or four days he suffered attacks of hysteria, a psychological disturbance that manifests itself in such abnormal and sometimes incapacitating physical symptoms as blindness, paralysis, or numbness and for which there appears to be no physical basis. He went back to duty, and during rifle practice a few days later, the symptoms reappeared. The army shipped him overseas on July 5. During grenade practice in France, a man in his platoon pulled the firing pin but panicked and let the grenade drop. Sergeant O. C. warned his men, and they all made it to safety. But that night after taps, his whole body began to shake, and he sweated profusely and felt numb in several parts of his body. Afterwards, he continued to complain that his whole body was trembling. The other soldiers began to call him "shakes." A doctor who examined him described him as "poor stuff." This sergeant was one of the more than six thousand cases admitted as hysteria victims.

Several men suffered from an illness called anxiety neurosis, which seemed to arise from a conflict between having to behave a certain way, particularly in battle, and the fact that this behavior threatened one's existence. It appeared chiefly in officers and in noncommissioned officers who had to stifle instinctive desires, like the urge to escape to safety, in order to lead others, though it also emerged in ordinary privates with a strong sense of responsibility. In training, these men learned to hide such emotions and, as Sergeant Blatt said, to "camouflage" their panic in combat. Some buried their feelings so deeply that they ceased to be aware of them; but prolonged stress followed by a traumatic occurrence on the battlefield, like a shell concussion, brought these conflicts to the surface and led to intense anxiety, depression, apathy, sleeplessness and nightmares, and shaking and a rapid pulse, with the soldier often oblivious to the cause of his suffering.

A. P., a Marine Corps private, was diagnosed as having anxiety neurosis. He had enlisted as an eighteen year old and, particularly after arriving in France, had come to relish military life. During March and April of 1918, he was under constant bombardment but recalled not being bothered by it. After a rest, he went back to the lines at Château-Thierry, welcoming the chance of getting some open warfare. During the first four days, he was under shell fire and saw a fair number of casualties but, nevertheless, remembered enjoying himself and did not seem afraid, although he wondered if one of the shells would "get" him. On June 5, his company advanced to relieve the French. He saw many dead Frenchmen, some with their heads blown off and others with their eyes staring at him. Though he had never been able to stand touching a corpse, his superiors sent him on a burial detail, which he found a horrifying and disgusting experience. It was only then that he grasped what shell fire was and what it did to people. Afterward, he could not sleep for several nights because the dead Frenchmen were constantly in his mind. Artillery attacks began to frighten him, and he trembled under fire but tried to conceal his fear and carry on. On June 14, while being shelled in the open, he started to shake, became weak, and went to the dressing station. Sent to the hospital, he had more terrifying dreams, and though he began to recover, he could not stand to look at the other patients who stammered and shook and jumped at the slightest noise.

A respondent to the World War I Survey, Duncan Kemerer, told how shell shock felt to a victim. Kemerer was a private in the 111th Infantry, Twenty-eighth Division. At Fismette, during a battle on the Vesle River, he was inside a stone building when enemy artillery began to attack. He remembered standing at the front door as the Germans made a direct hit, blowing him about fifteen to twenty feet from the building. He stumbled to his feet and began to run toward the river, without helmet, rifle, or gas mask and with his shredded clothing splattered with blood from a wound in the back of his head. "The soldier in front of me collapsed from shrapnel," he wrote, "and with the flash and the noise and my weakened condition, without food or water for two to three days, I became shell-shocked, losing consciousness." When he awoke, he was lying in the cellar of a company headquarters on the other side of the river awaiting evacuation. Each time a shell burst overhead, he yelled and tried to protect himself by digging a hole with his fingers in the dirt floor. Then he lost consciousness again.

Kemerer was sent to Base Hospital no. 15 in Chaumont, where, whenever he heard something drop, a spoon, for instance, or when there was an unusually loud noise, he screamed and buried himself beneath the sheets or under the bed. But with "good care from French nurses, maids and doctors plus rest and good food," he was considered well enough to return to his company and endured more harrowing experiences, including being blown off his feet by an enemy shell during a rout.

When men like Kemerer became shell-shock casualties, they encountered a system that aimed to arrest or cure their illnesses but that had the overriding objective of getting them back into battle. According to army medical procedures, which could not always be carried out because of battle or logistical conditions, a psychiatrist at the division triage immediately behind the battle line evaluated "nervous" cases, sent those who were fit directly back to their units, and arranged to rest, feed, warm, and provide basic psychological care to certain others, particularly exhaustion cases. The remainder were evacuated to a nearby field hospital close to the front if they could be administered to there or to a special division neurological hospital further back if no field hospital could receive and care for them. Every effort was made to treat these men before their symptoms became fixed and to avoid separating them from their units; for if they went too far to the rear, it became extremely difficult to salvage them for further front-line duty. Psychiatrists tried to forestall what they called secondary gain, which for shell-shocked soldiers meant an advantage of safety and freedom from hardship that could turn them into chronic casualties. They also hoped to prevent epidemics of shell shock at the front.

The prescribed way to handle these men was to treat them simply and expeditiously, preserving as far as possible their emotional ties to comrades on the battle line. Thus, in the First Division an officer of the day admitted "nervous" casualties to the division neurological hospital, told them in uncomplicated fashion what ailed them, and reassured them about the prognosis. The soldiers were then bathed, fed, and put to bed, where they usually fell into a deep sleep that lasted from thirty-six to forty-eight hours. When they awoke, the ward psychiatrist talked with each patient, explaining his condition and treating his symptoms with suggestion or persuasion; for instance, he described to patients tortured by battle nightmares how dreams worked and urged them to ventilate their emotion-charged experiences during waking hours. Later, the hospital's commanding officer saw each patient privately and discussed with him the history of his problem, interpreted his symptoms in a way intended to remove any residue of mystery or horror, and offered reassuring suggestions. Then, after three or four days, the patients were placed on a schedule that included exercise, calisthenics, and recreation.

Though the chief surgeon of the AEF claimed that even serious neuropsychiatric cases could be returned to duty, including men with severe concussions who had been blown up by shells or buried in trenches, results of treatment varied considerably. The type of disorder played a part. Psychiatrists found that symptoms developed in the aftermath of a definite trauma and followed by a lapse of consciousness were relatively easy to relieve, while anxiety cases proved much less tractable. Some men were considered fit to go back to combat, others were assigned to temporary noncombat duty or sent for the duration to the Services of Supply. Very difficult cases were hospitalized in Europe or returned to the United States.

Army physicians felt it was especially hard to cure patients not treated early, chronic cases, and soldiers with a certain type of intellect. In his account of World War I neuropsychiatry that appears in the official Medical Department history, Dr. Edward A. Strecker, the Twenty-eighth Division psychiatrist, described the problem presented by those in this last category:

> The intellectual status of the patient was not without its effect. The relatively ignorant soldier was usually softer clay in the physician's hands than was the one in whom learning and training had sharpened the habit of questioning, scrutinizing and weighing in the balance.

Army psychiatrists believed a man's education or his values might predispose him to certain types of neuroses. They thought soldiers with some education and those with a "well-developed ethical sense" would be especially likely to suffer from a conflict between the need to save themselves and a desire to do their prescribed duties. The resulting anxiety could make an officer incapable of acting.

Medical personnel believed some shell-shock casualties were trying to avoid combat by manipulating the treatment system, feigning or using shell-shock symptoms to avoid combat. At the same time, medical personnel manipulated the psychiatric casualties, genuine or otherwise, starting with their first encounters. They did this partly by controlling language. If an enlisted sanitary corpsman with no knowledge of how to diagnose put the words "shell shock" on a man's medical tag, the patient clung to that label regardless of what the psychiatrist said. The army, therefore, decreed that corpsmen must never label soldiers they picked up on the field or encountered at dressing stations as shell-shock casualties. The mandatory term was "N.Y.D. (Nervous)."* This designation did not sound like the kind of illness that allowed a man to leave the field with honor. It gave the soldier nothing to cling to and kept him open to the suggestion of medical officers that he was only tired and a little nervous and would be fit for duty after a short rest. So eager was the army to abolish the expression "shell shock" that it eliminated the phrase from official communications; but it could not stop soldiers from using it in ordinary speech.

To prevent war neurosis patients from lapsing into a civilian state of mind, they were treated in a military atmosphere, wore their uniforms whenever possible, and performed military duties. Medical workers were indoctrinated to take a positive view of all cases and to convey to patients the feeling that they were bound to get better and would soon go back into battle. Doctors and nurses stressed the glory and traditions of the division, regiment, and company and the important part each soldier played in his unit. Chaplains delivered "wisely and cleverly designed sermons," to use

*Meaning, "Not yet diagnosed."

the words of the official Medical Department history, "touching on the spiritual phase of courage, loyalty, devotion, and patriotism." Weekly patriotic talks by staff members, war posters that decorated recreation huts, bulletins announcing the exploits of particular gallant units and of individual soldiers, discussions of military citations, and selected gossip from the front kept the war in the thoughts of patients and reinforced the idea that they were preparing to return to battle. During the Meuse-Argonne offensive, soldiers were brought out to watch columns of enemy prisoners, a sight that, an army psychiatrist observed, evoked satisfaction and even patriotic fervor.

At the same time, army medical personnel discouraged thoughts of further evacuation, suggesting that to leave one's unit meant giving up its honors and rewards and abandoning paternal relationships with one's officers and fraternal ties to fellow soldiers, which might cause one to end up a lonely, isolated soldier. All this psychological manipulation, a division psychiatrist observed, was carefully thought out, offering emotional carrots, as he called them, or wielding emotional sticks according to the type and amount of "stimulation" required.

Two groups appeared to need special stimulation. One consisted of men who seemed to have recovered and might even have said they wanted to return to their regiments but somehow could not take the final step. The doctors tried to reason with them and appealed to them individually; but if that did not work, they were gathered in a tent and given what Dr. Strecker called an informal talk. The speaker began by describing the situation at the front, telling them what the division and their own units were doing. He emphasized "the acute need for every available man and the fact that comrades were suffering because of their absence." Finally, he made a "dramatic request" for them to volunteer for immediate service. The result, Strecker said, was "always highly gratifying. . . . "

A second group, though free of major symptoms, seemed willfully uncooperative. The army might have threatened to court-martial them if they refused to return to the lines, but it chose not to—a wise decision in Strecker's view, because forcing them back into battle might not have worked for very long and was ethically questionable. Instead, the medical workers followed a less direct approach. Strecker recalled:

> Although such men were not treated with undue severity, or with any malice, they soon found that an invisible barrier had been erected between them and the other patients. They were denied certain privileges and had to do most of the distasteful work, such as policing the grounds, digging latrines, and the like. No one was permitted to impugn their motives, yet on every side they were confronted by a questioning attitude. Always the op-

portunity was afforded them, and indirectly encouraged, to talk over the situation with one of the physicians; always there was the invitation and the temptation to change their status to a happier and more honorable one. About 90 percent of this group were eventually reached by such a simple method.

Military psychiatrists used suggestion and a variety of tricks to treat patients who might be trying to deceive the medical-treatment system. In amnesia cases, the physician asked the patient a series of questions to determine how real and how serious his loss of memory was. If a patient was shaking, the doctor employed a technique, called passive relaxation of flexion and tension, accompanied by suggestion to induce him to stop and to show him that his motion could be halted. When a limb seemed paralyzed, the psychiatrist moved it and gradually let it go to see if the patient was able and willing to control it by himself; if this did not work, he might administer an electric shock to demonstrate that it could be moved. Soldiers with obstinate tremors or convulsive movements were sent into places where they thought no one could see them. This, the official history says, had a "wholesome" effect on the symptoms. A patient whose problems seemed intractable would be placed briefly in the midst of a small group of soldiers who had recovered and were waiting for transportation to the front.

From time to time, a psychiatrist tried to encourage recalcitrant patients to improve by quietly arranging for someone who had made a particularly striking recovery to behave a certain way in front of them. Physicians also made it possible for so-called chronic cases to observe the removal of symptoms in another patient. The psychiatrist might plan to have a recalcitrant soldier overhear his comments, seemingly by chance, or he might stage elaborate consultations for their psychic effect alone. To treat a patient with "troublesome" symptoms, he sometimes aroused the man's curiosity, then kept postponing the "final séance."

What were the consequences of the war for the men who were shellshocked—those treated and those who were not? Army physicians claimed a substantial number of cures for patients treated properly. Yet some of those who went back to the lines after a short period of rest and care returned with psychiatric problems, and some of those who were treated suffered more serious breakdowns afterward. The mental health of soldiers who subsequently died in battle is almost impossible to ascertain, and little is known about the postwar condition of surviving doughboys who developed psychoneurotic symptoms, whether treated or not. So the long-term outcome for treated AEF shell-shock patients is uncertain.

Two postwar studies, one in 1919–1920 and the second in 1924–1925, each based on slightly more than 750 responses from the 2,590 men who had been treated at Base Hospital no. 117, showed a substantial percentage of lingering mental problems (see Tables 2 and 3). These studies do not describe the general experience of shell-shocked AEF soldiers, for many of those who developed the symptoms of the disease were never treated or did not receive appropriate care for years; nor do the studies of men seen in Base Hospital no. 117 represent the typical experience of AEF veterans who had been treated during the war for psychoneurotic symptoms, for only men who could not be administered to near the front were supposed to be evacuated to base hospitals. In addition, the surveys of former patients of Base Hospital no. 117 were not very rigorous.

Yet even if there were no carefully designed and well-executed studies of veterans with shell shock, it is known that thousands of men remained in veterans hospitals with lingering effects of mental illness related to the war. At the beginning of 1922, there were some nine thousand patients in veterans hospitals suffering from neuropsychiatric disease that the government connected to military service. Two and a half years later, the number

Table 2 Distribution of Cases of Former Patients in 1919–1920 Follow-up

Condition	Percent
Normal	38.9
Neurotic	22
Fatigued	17.3
Disabled	20.4
Psychotic	1.3

Source: Pearce Bailey, et al., *Neuropsychiatry* (Washington, D.C.: GPO), 449.

Note: Normal men were those who adjusted to civilian life well and who were normally healthy and happy, though some noted a tendency to become angry or excited easily, be a little nervous, restless, or forgetful, have occasional slight headaches or dizziness or other "mild" neurotic symptoms.

Neurotic veterans had made partial readjustments but continued to suffer from one or more "rather severe" nervous problems, including tics, insomnia, weakness, speech defects, jumpiness, memory disorders, severely handicapping periodic mood disorders, and inability to concentrate.

Fatigued men could not work regularly without suffering and being confined to bed. They tired easily and suffered from severe headaches, lack of ambition, and depression.

Disabled men were hospitalized for a psychoneurotic or physical disease.

Table 3   Distribution of Cases of Former
Patients in 1924-1925 Follow-up

Conditions	Percent
Normal	36.9
Neurotic	43.9
Fatigued	9.8
Disabled	7.7
Psychotic	1.7

Source: Pearce Bailey, et al., *Neuropsychiatry* (Wash-
ington, D.C.: GPO), 466.
Note: definitions of conditions are the same as for
Table 2.

had increased by almost a thousand, while the total of all patients with
service-connected problems dropped from about 31,000 to about 23,000.
As late as June 1940, out of 11,501 World War I veterans in hospital or
domiciliary care for service-connected illnesses, 9,305 were neuropsychiatric cases. Thousands of other men were treated privately for the
aftermath of shell shock or simply endured it.

The World War I Survey and other records gathered after the war
suggest what happened to some of these casualties. They tell how, decades
after the Armistice, Frank Di Nino, who had carried wounded men at
Saint-Mihiel and Château-Thierry, had a flashback about the horrors of
the trenches and described it to his wife. This was the first time, she said,
he had spoken a word to her about the war. Private First Class Harvey L.
Hendricksen wrote that he was compelled to relive in "mirages" the times
when he had narrowly escaped from death, while Duncan Kemerer, the
nineteen-year-old private blown out of a building at Fismette, mentioned
when he was seventy-one that he still suffered from shell shock.

Several of these veterans reported problems in adjusting to civilian life
that may have arisen from or been brought to the surface by the trauma of
war. Samuel Z. Orgel, a lieutenant colonel in the Forty-second Division,
recalled being mentally exhausted during the war but not allowed to rest.
It took him some years, he said, to get to a "normal position." Private First
Class Stephen R. Barrett, a male nurse who had enlisted in the Army
Medical Service in a patriotic spirit and intending to apply his skills at the
front, spent many days in combat. After discharge, he found himself "very
unstable," unable to organize himself or stay put. He had planned to study
medicine but never went back to school. Service in the army, he stated,
had "wrecked my future."

Malcolm Aitken remembered coming out of the U.S. Marine Corps a
shadow of his former self—physically, with amoebic dysentery; and psy-

chically, with "mental reaction to combat; night-mares, etc." Books about the World War I battlefields (*Through the Wheat,* which appeared five years after the Armistice, and *All Quiet on the Western Front,* published eleven years after the fighting ended) gave Aitken terrifying dreams. Sergeant Ralph T. Williams, a private first class in the Second Division, told how he was not "readjusted" enough when he was discharged to go back to school and took until 1924 to get established. Walter Zukowski of the Thirty-second Division stated he had been in a "shell-shock condition" for three years after returning to America. He suspected something was wrong and managed to conceal his "strange behavior." But at night, he was "still fighting the war."

Twenty years after the fighting ended, former Lieutenant Joseph Lawrence met several veterans of his company. A few were doing well, he said, and some only fairly so. Many were "derelicts" who could not readjust from the "shock of war." Several of these men could have ended up derelicts whether they had joined the AEF or not. The war precipitated mental disorders in others that might have appeared if they had never left civilian life. Still, in weighing the gains and losses of America's Great Crusade, we should add the psychic casualties of shell shock to the number of victims of the First World War.

While the war's long-term effects on AEF psychiatric casualties were somewhat uncertain, it is clear that U.S. participation in World War I benefited American psychiatrists. Adapting techniques that Allied physicians had devised, the medical people who treated shell-shock victims made a significant contribution to AEF military power by returning men to fight and probably by forestalling shell-shock epidemics in the front lines. Despite difficulties in carrying out their program in wartime chaos, often on the edge of the battlefield, they demonstrated how psychiatric methods used to treat individual patients could serve an important social objective—winning the World War. At a time when psychiatry was seeking to demonstrate both its scientific qualities and its ability to cure, the war enabled its military practitioners to produce tangible though perhaps short-term results. They created and carried out a vast mental-hygiene plan that indicated what might be done in peacetime for the people of the United States. Finally, army psychiatrists had the satisfaction of easing, at least for a while, the suffering of soldiers—even that of those who returned to the stress of the battleground—and keeping some of them from becoming lifetime mental casualties.

The relationship of these psychiatrists with doughboy patients was elitist and paternalistic. In the official history, Dr. Strecker and his colleagues made no effort to hide how they thought about the young men they treated. They were "clay" to be "molded" and, if necessary, mystified;

for, as Strecker said, "some of the more complex forms of technique depend largely for their suggestive value on the veil of mystery which surrounds them." This position was not at all unusual at the time in a profession in which practitioners often assumed a god-like attitude that they felt was an aid to recovery. Not only did therapists believe this stance promoted healing by building confidence in their powers, but some of them held that successful treatment required the patient to reenact childhood dependence, with the psychiatrist assuming a paternal role. For instance, Dr. Abram Kardiner, who treated numerous World War I veterans, remarked that one of his patients "never assumed an attitude of dependence . . . toward the physician . . . and hence anything in the nature of a transference [a process involved in psychotherapy] was impossible." In an army treatment center, where psychiatrists determined whether a soldier had to return to the battleground or not, they probably appeared even more parent-like or god-like than usual.

Psychiatrists were not the only medical men who viewed young Americans of military age from an elite perspective. A Cleveland physician who examined potential draftees and would-be draft avoiders wrote: "With few exceptions, these men lost all guile with their clothes. They appeared helpless and could, with a little tact, be handled like school children." And one did not have to be a physician to hold this attitude. Recall the CTCA lecturer who hammered the virtues of social purity into soldiers until they were almost as "pliable as putty" (see p. 102). One of the paradoxes of America's war for democracy is that so many of the people who participated in it, whether conservative businessmen, liberal politicians, or social reformers, thought of themselves as an elite and acted as such, trying to shape the attitudes and behavior of common men and women.

The treatment of so-called shell-shock casualties in the American Expeditionary Force exemplified the style of management often used by the United States government to fight the Great War—the bargaining and the manipulation of one party by another, the efforts of authorities to control behavior by controlling language, and the use of programs that fostered the welfare of particular groups (in this case, mental patients) as a way to accomplish the government's objectives. In the psychoneurological treatment center, as elsewhere in America, those in charge used exhortation, suggestion, and indirection to induce people to do voluntarily what their leaders believed had to be done.

Epilogue

Early in 1918, President Wilson announced the set of official American war aims that Walter Lippmann and other members of the Inquiry had helped prepare—the Fourteen Points. These included such objectives as the removal of barriers to international commerce, arms reduction, boundary adjustments that provided ethnic groups with nations of their own, political self-determination, and a league to guarantee international stability. Wilson knew that the Allies had interests that diverged from those of the United States and that they intended to carry out territorial agreements made in secret during the war. Nevertheless, he hoped that America's dominant financial power and its military contribution would induce them to accept his objectives and, together with the United States, establish a secure world system based on his principles.

Although the president was able to achieve some of his goals, the peace settlements proved highly unstable. They left Germany badly wounded and full of bitterness—yet still capable of preparing for another great war—France and England spiritually and physically enfeebled, and Europe as a whole economically weak and threatened by revolution. The victors agreed to fulfill part of the president's vision by creating the League of Nations, but despite all his efforts, the Congress of the United States refused to approve U.S. membership.

Wilson prophesied that failure to join the league would lead to disastrous consequences for America. It would mean that the nation would have to be constantly ready for war, with a great standing army, vast supplies of weapons, secret plans, and a spy (or, to use what Wilson called the more polite term, "intelligence") system. In these circumstances, free debate and social reform could not take place because everyone would be under orders from the government. It was impossible for a country to be both militarized and free.

As if to defy his own long-range vision, Wilson's government demobilized most of the American armed forces in the year after the Armistice and dissolved the domestic infrastructure that had supported them. A few parts of the wartime system lasted for a while. The Department of Commerce continued to encourage cooperation between businesses. The United States Shipping Board and the War Finance Corporation aided American companies, while the Veterans Bureau assisted former servicemen. More significant, ideas about the way government should act in a crisis persisted, ready to be used in the next emergency—the Great Depression.

The occupant of the White House when that emergency began had mixed feelings about applying those lessons. President Hoover followed wartime examples by attempting to influence the public with personal exhortation and by encouraging volunteer activities. He told Americans that if they gave the same service and the same confidence to their government and institutions and displayed the unity and solidarity they had shown during the Great War, the depression could be overcome. Like the members of the Council of National Defense in the early months of the war, he encouraged business leaders to engage in a cooperative attack on the nation's economic problems. He re-created the War Finance Corporation as the Reconstruction Finance Corporation, which supplied credit to railroads, insurance companies, banks, and other weakened financial institutions and made loans to the states for public-works relief projects. Meanwhile, Hoover's Federal Farm Board established public corporations, similar to the ones the wartime Food Administration had created, to stabilize commodity prices.

Hoover used these federal agencies reluctantly, partly because he shared Wilson's qualms about the effects on American society of the kind of management that had operated during the war—a centralized despotism, he called it. He feared that to continue that kind of regimentation in time of peace would destroy the freedoms of the American people and wreck the private-enterprise system. But his successor, who had also served in the Wilson government, as assistant secretary of the navy, had fewer qualms about employing wartime techniques. Indeed, Franklin Roosevelt's New Deal, especially in its earliest phase, resurrected much of the war welfare state. The New Deal's Civilian Conservation Corps, which assembled unemployed youths in army recruiting centers and put them to work under military supervision, was a peacetime version of World War I army camps, complete with uniforms, tents, and military bugle calls. The earliest New Deal public-housing venture descended directly from the wartime federal housing agencies. Its first director had served during the war as chief of production in the housing division of the United States Shipping Board. The Tennessee Valley Authority, one of the most successful New Deal projects, originated with a hydroelectric power plant the government had constructed to produce explosives for the First World War.

New Deal agriculture, labor, and welfare programs were all related in some way to earlier war activities. The Agricultural Adjustment Administration, headed by a former member of the War Industries Board, George Peek, operated a variation of the wartime wheat program. While the government had raised the output and limited the price of wheat during the war, the AAA limited the output of commodities and raised their prices. The National Labor Relations Board, created to give labor some protection in the new mobilization of the economy, followed basic principles of the War Labor Board. The Federal Emergency Relief Administration and the Social Security Administration were part of a line of development that extended back to the Great War, when the Wilsonians made payments to veterans and the families of servicemen. New Dealers extended government social insurance to cover new groups of civilians: the old, the disabled, widows, orphans, and the unemployed.

There were direct links between the War Industries Board and the National Recovery Administration, the most important of the agencies with which the New Deal first tried to manage industrial recovery. Several NRA staff members and its director, ebullient General Hugh Johnson, were War Industries Board veterans. Like the WIB, the NRA brought businessmen and government officials together and gave them the task of regulating American industry. It assembled trade association representatives in Washington, encouraged competing companies to make agreements for regulating competition, and saw to it that the antitrust laws were not enforced. The War Industries Board had helped set prices to secure maximum output of military goods. The NRA tried to raise prices to the point at which businessmen would reopen factories and start producing again for civilian markets. Both agencies pursued their goals through centralized yet largely indirect manipulation of the economy.

Like the war welfare state, the New Deal used intellectuals to formulate and carry out its policies, and like their predecessors during the First World War, the New Deal intelligentsia regarded the crisis of the Great Depression as an opportunity to test their ideas and turn them into government programs. New Dealers tinkered constantly, changing or abandoning programs, building one organization on top of another, acting much as the Wilson administration had acted in the early, indecisive state of mobilization.

The people who operated Roosevelt's programs, like the wartime Wilsonians, mixed incentives with compulsion—hedging farmers and businessmen about with restrictive laws and bureaucratic regulations while providing contracts and loans to businessmen, subsidizing farmers, creating jobs, making welfare payments, and constructing bridges, libraries, dams, post offices, and other widely appreciated public works. Though these programs provided something for almost everybody, the New Dealers tended, like the wartime government, to respond most generously to the most powerful, best-organized interest groups. Large growers—not small farmers, sharecroppers, and hired hands—benefited most from the AAA. Multimillion dollar corporations, not small businessmen, dominated the National Recovery Administration. Usually the strongest unions made the largest advances during the early New Deal years, and not until old people began to organize behind Dr. Francis Townsend's pension program and threatened to become a significant pressure group did the administration push hard for Social Security. Some critics on the left distrusted this inclination to dispense the largest rewards to the strongest interest groups, seeing an ominous similarity between the New Deal and Italian Fascism. But the actual model was not the corporate state of Benito Mussolini; it was the wartime government of Woodrow Wilson.

Warlike rhetoric and attitude management methods similar to those used in the Great War played important parts in the early Roosevelt administrations. New Dealers recognized that much of the nation's economic problem was really psychological—defeatism, disunity, failure of confidence—and they drew on wartime mass-psychological techniques to lead the country toward recovery. Roosevelt described himself, in his first inaugural address, as leader of a "great army of our people, dedicated to a disciplined attack upon our common problems." He pledged that, if other measures failed, he would go to Congress for "power to wage a war against the emergency as great as the power that would be given . . . if we were in fact invaded by a foreign foe." After this call to arms, the National Recovery Administration began to publicize itself with rallies and parades reminiscent of the Liberty Bond campaigns. New Deal agencies described their exploits and constructed favorable images of the government through press releases, feature articles, pamphlets, and motion pictures; and the president began a series of radio broadcasts to influence public opinion.

Along with these elements of continuity between the New Deal and the war welfare state, there were important differences. Many of the economic problems New Dealers faced were quite dissimilar to those of their wartime predecessors. For instance, the New Deal tried to combat falling prices, not inflation, and its job was to start the economy working again, not to convert an already booming economy from civilian to military production. The welfare state that the New Dealers built proved considerably more durable than its predecessor. Wilson relied on dollar-a-year men, who after the Armistice returned to private employment. Roosevelt constructed a standing bureaucracy. During the Depression years, the federal government had no central propaganda agency like the Creel Committee. It was considerably more tolerant of dissent from the left and showed almost no interest at all in managing the morals of the American people. Unlike Wilson's war administration, the New Deal, particularly after 1934, projected an image of hostility toward big business and, instead of building public esteem for businessmen as wartime agencies had done, occasionally attacked and always overshadowed them. Even before the Supreme Court finished off the National Recovery Administration by declaring it unconstitutional, New Dealers had begun to reject the principle of business self-regulation, a fundamental precept of the wartime management system.

Still, the centrally managed society of World War I both presaged and contributed to the rise of federal power in the 1930s. And it also foreshadowed much of what happened in the decades after the New Deal as the United States entered a long era, as Wilson had foreseen, of warfare and continuous preparation for war. In those years, there was a general elaboration of the war welfare state, in which the central government attempted to influence the way Americans felt about external and internal threats and provided countless citizens—from the very rich to the very poor—and numerous businesses, professions, and institutions with a stake in the national security system. Thus, the history of America in the Great War provides a glimpse of what America would become during most of what remained of the twentieth century.

Appendix

Following are excerpts from the Espionage Act of June 15, 1917, the Trading with the Enemy Act of October 6, 1917, and the Sedition Act of May 16, 1918.°

Espionage Act, Title I, Section 3

Whoever, when the United States is at war, shall willfully make or convey false reports or false statements with intent to interfere with the operation or success of the military or naval forces of the United States or to promote the success of its enemies and whoever, when the United States is at war, shall willfully cause or attempt to cause insubordination, disloyalty, mutiny, or refusal of duty, in the military or naval forces of the United States, or shall willfully obstruct the recruiting or enlistment service of the United States, to the injury of the service or of the United States, shall be punished by a fine of not more than $10,000 or imprisonment for not more than twenty years, or both.

Espionage Act, Title XII, Sections 1, 2, and 3

Every letter, writing, circular, postal card, picture, print, engraving, photograph, newspaper, pamphlet, book, or other publication, matter, or thing, of any kind, in violation of any of the provisions of this Act is hereby declared to be nonmailable matter and shall not be conveyed in the mails or delivered from any post office or by any letter carrier: *Provided*, That nothing in this Act shall be so construed as to authorize any person other than an employe of the Dead Letter Office, duly authorized thereto, or other person upon a search warrant authorized by law, to open any letter not addressed to himself.

°Source: *U.S. Statutes at Large*, vol. 40, part 1.

Every letter, writing, circular, postal card, picture, print, engraving, photograph, newspaper, pamphlet, book, or other publication, matter, or thing, of any kind, containing any matter advocating or urging treason, insurrection, or forcible resistance to any law of the United States, is hereby declared to be nonmailable.

Whoever shall use or attempt to use the mails or Postal Service of the United States for the transmission of any matter declared by this title to be nonmailable, shall be fined not more than $5,000 or imprisoned not more than five years, or both. Any person violating any provision of this title may be tried and punished either in the district in which the unlawful matter or publication was mailed, or to which it was carried by mail for delivery according to the direction thereon, or in which it was caused to be delivered by mail to the person to whom it was addressed.

Trading with the Enemy Act, Section 19 (excerpt)

That ten days after the approval of this Act and until the end of the war, it shall be unlawful for any person, firm, corporation, or association, to print, publish, or circulate, or cause to be printed, published, or circulated in any foreign language, any news item, editorial or other printed matter, respecting the Government of the United States, or of any nation engaged in the present war, its policies, international relations, the state or conduct of the war, or any matter relating thereto: *Provided*, That this section shall not apply to any print, newspaper, or publication where the publisher or distributor thereof, on or before offering the same for mailing, or in any manner distributing it to the public, has filed with the postmaster at the place of publication, in the form of an affidavit, a true and complete translation of the entire article containing such matter proposed to be published in such print, newspaper, or publication, and has caused to be printed, in plain type in the English language, at the head of each such item, editorial, or other matter, on each copy of such print, newspaper, or publication, the words "True translation filed with the postmaster at   on   (naming the post office where the translation was filed, and the date of filing thereof) as required by the Act of   (here giving the date of this Act).

Any print, newspaper, or publication in any foreign language which does not conform to the provisions of this section is hereby declared to be nonmailable, and it shall be unlawful for any person, firm, corporation, or association, to transport, carry, or otherwise publish or distribute the same, or to transport, carry or otherwise publish or distribute any matter which is made nonmailable by the provisions of the Act relating to espionage, approved June fifteenth, nineteen hundred and seventeen. . .

The Sedition Act of May 16, 1918

Be it enacted by the Senate and House of Representatives of the United States of America in Congress assembled, That section three of title one of the Act entitled "An Act to punish acts of interference with the foreign relations, the neutrality, and the foreign commerce of the United States, to punish espionage, and better to enforce the criminal laws of the United States, and for other purposes," approved June fifteenth, nineteen hundred and seventeen, be, and the same is hereby amended so as to read as follows:

"Sec. 3. Whoever, when the United States is at war, shall willfully make or convey false reports or false statements with intent to interfere with the operation or success of the military or naval forces of the United States, or to promote the success of its enemies, or shall willfully make or convey false reports or false statements, or say or do anything except by way of bona fide and not disloyal advice to an investor or investors, with intent to obstruct the sale by the United States of bonds or other securities of the United States or the making of loans by or to the United States, and whoever, when the United States is at war, shall willfully cause or attempt to cause, or incite or attempt to incite, insubordination, disloyalty, mutiny, or refusal of duty, in the military or naval forces of the United States, or shall willfully obstruct or attempt to obstruct the recruiting or enlistment service of the United States, and whoever, when the United States is at war, shall willfully utter, print, write, or publish any disloyal, profane, scurrilous, or abusive language about the form of government of the United States, or the Constitution of the United States, or the military or naval forces of the United States, or the flag of the United States, or the uniform of the Army or Navy of the United States, or any language intended to bring the form of government of the United States, or the Constitution of the United States, or the military or naval forces of the United States, or the flag of the United States, or the uniform of the Army or Navy of the United States into contempt, scorn, contumely, or disrepute, or shall willfully utter, print, write, or publish any language intended to incite, provoke, or encourage resistance to the United States, or to promote the cause of its enemies, or shall willfully display the flag of any foreign enemy, or shall willfully by utterance, writing, printing, publication, or language spoken, urge, incite, or advocate any curtailment of production in this country of any thing or things, product or products, necessary or essential to the prosecution of the war in which the United States may be engaged, with intent by such curtailment to cripple or hinder the United States in the prosecution of the war, and whoever shall willfully advocate, teach, defend, or suggest the doing of any of the acts or things in this section enumerated, and whoever shall by word or act support or favor the cause of any country with which the United States is at war or by word or act oppose the cause of the United States therein, shall be punished by a fine of not more than $10,000 or imprisonment for not more than twenty years, or both: *Provided,* That any employee or official of the United States Government who commits any disloyal act or utters any unpatriotic or disloyal language, or who, in an abusive and violent manner criticizes the Army or Navy or the flag of the United States shall be at once dismissed from the service. Any such employee shall be dismissed by the head of the department in which the employee may be engaged, and any such official shall be dismissed by the authority having power to appoint a successor to the dismissed official."

Essay on Sources

Most of the sources on which this book is based are catalogued in two bibliographies, David R. Woodward and Robert Franklin Maddox, *America and World War I: A Selected Annotated Bibliography of English-Language Sources* (New York, 1985); and Ronald Schaffer, *The United States in World War I: A Selected Bibliography* (Santa Barbara, Calif., 1978). These works contain numerous references to matters related to the subject of this book but not explored here, for instance, the origins of the war, United States prewar and wartime diplomatic relations with the belligerents, the peace settlement, and American reactions to the Russian Revolution. The pages that follow describe the published and unpublished sources used to write *America in the Great War* and constitute a brief, highly selective guide to works on the management of American society during that conflict.

Introduction

Two recent comprehensive studies of America in the years of the First World War are Robert H. Ferrell, *Woodrow Wilson and World War I: 1917–1921* (New York, 1985), and David M. Kennedy, *Over Here: The First World War and American Society* (New York, 1980), both lucidly written and based on extensive research in unpublished as well as published sources.

On Woodrow Wilson, the reader should consult *The Papers of Woodrow Wilson,* edited by Arthur Link et al. (Princeton, N.J., 1946–), and Link's multivolume biography, *Wilson* (Princeton, N.J., 1947–), which has reached the point where Wilson decides for war. Edwin A. Weinstein, *Woodrow Wilson: A Medical and Psychological Biography* (Princeton, N.J., 1987), and Alexander L. and Juliette George, *Woodrow Wilson and Colonel House: A Personality Study* (New York, 1964), make insightful connections between Wilson's policies and his character and (in the case of Weinstein) the physical condition of Wilson's brain. John M. Mulder,

Woodrow Wilson: The Years of Preparation (Princeton, N.J., 1978), interprets the religious context of the future president's ideas and policies. A thoughtful and, on the whole, highly sympathetic study of Wilson appears in John Milton Cooper, *The Warrior and the Priest: Woodrow Wilson and Theodore Roosevelt* (Cambridge, Mass., 1983).

For Wilson's foreign policy and the decision for war in 1917, see, in addition to the works listed above, Frederick S. Calhoun, *Power and Principle: Armed Intervention in Wilsonian Foreign Policy* (Kent, Ohio, 1986), which occasionally sees as principle what others would regard as rationalization; Lloyd Gardner, *Safe for Democracy: The Anglo-American Response to Revolution, 1913–1923* (New York, 1984); and Friedrich Katz, *The Secret War in Mexico: Europe, the United States and the Mexican Revolution* (Chicago, 1981). All of these books illuminate Wilson's policy toward the European belligerents by explaining how he dealt with other nations. Martin J. Sklar, "Woodrow Wilson and the Political Economy of Modern United States Liberalism," in *For A New America: Essays in History and Politics from Studies on the Left, 1959–1967*, edited by James Weinstein and David W. Eakins (New York, 1970), 46–100, discusses in a critical way the economic conceptions that to a significant extent underlay Wilson's view of foreign affairs, while Edward H. Buehrig, *Woodrow Wilson and the Balance of Power* (Bloomington, Ind., 1955), examines the impact of national security considerations on the president and his advisors.

Chapter 1. Managing American Minds

Two books tell the history of the Committee on Public Information: Stephen Vaughn, *Holding Fast the Inner Lines: Democracy, Nationalism, and the Committee on Public Information* (Chapel Hill, 1980), and James R. Mock and Cedric Larson, *Words that Won the War: The Story of the Committee on Public Information* (Princeton, N.J., 1939). George Creel offers an accounting in *How We Advertised America* (New York, 1920) and in United States Committee on Public Information, *The Creel Report: Complete Report of the Chairman of the Committee on Public Information 1917:1918:1919* (1920; reprint, New York, 1972).

One of the best ways to understand the Wilson administration's efforts to reach the minds of the American people is to read Committee on Public Information publications, including its "War Information Series" pamphlets, "Loyalty Leaflets," "National School Service" newsletter, and "Red, White and Blue" series. CPI posters are reproduced in such published collections and catalogs as Maurice Rickards, *Posters of the First World War* (New York, 1968), and the George C. Marshall Research

Foundation's *Posters of World War I and World War II in the George C. Marshall Foundation* (Lexington, Va., 1979). Most of the posters discussed in this chapter are in the collections of the New York Public Library and the Hoover Institution Archives, Stanford, Calif. Michele Shover analyzed the way the CPI and other agencies used pictures of women in "Roles and Images of Women in World War I Propaganda," *Politics and Society* 5, no. 4 (1975): 469–86. For analyses of film propaganda, see Larry Wayne Ward, *The Motion Picture Goes to War: The U.S. Government Film Effort during World War I* (Ann Arbor, Mich., 1985), and Craig W. Campbell, *Reel America and World War I* (Jefferson, N.C., 1985).

The government's difficulties in managing public opinion entirely are suggested in Christopher Gibbs, *The Great Silent Majority: Missouri's Resistance to World War I* (Columbia, Mo., 1988); Gerald R. Gill, "Afro-American Opposition to the United States' Wars of the Twentieth Century: Dissent, Discontent and Disinterest," Ph.D. diss., Howard Univ., 1985; Theodore Kornweibel, Jr., "Black America's Negative Responses to World War I," *South Atlantic Quarterly* 80 (Summer 1981): 322–38; and James Weinstein, *The Corporate Ideal in the Liberal State: 1900–1918* (Boston, 1969), Chapter 8.

Chapter 2. Controlling Dissent

Accounts of war hysteria, private and public, appear in Zechariah Chafee, Jr., *Freedom of Speech* (New York, 1920), William Preston, Jr., *Aliens and Dissenters: Federal Suppression of Radicals, 1903–1933* (Cambridge, Mass., 1963), Donald Johnson, *The Challenge to American Freedoms: World War I and The Rise of the American Civil Liberties Union* (Lexington, Ky., 1963), and Horace C. Peterson and Gilbert C. Fite, *Opponents of War, 1917–1918* (Madison, Wis., 1957). Frederick C. Luebke, *Bonds of Loyalty: German-Americans and World War I* (DeKalb, Ill., 1974), provides an evenhanded, chilling account of the causes and effects of hostility toward that large minority group. Herbert Shapiro, "The Herbert Bigelow Case: A Test of Free Speech in Wartime," *Ohio History* 81 (Spring 1972): 108–21, tells what happened to a prominent antiwar minister. Recent works that discuss the impact of the war on the courts and the Constitution are Richard Polenberg, *Fighting Faiths: The Abrams Case, the Supreme Court, and Free Speech* (New York, 1987), and Paul L. Murphy, *World War I and the Origins of Civil Liberties in the United States* (New York, 1979).

The activities of state councils of defense, some more reactionary and politicized than others, can be traced in William J. Breen, *Uncle Sam at Home: Civilian Mobilization, Wartime Federalism, and the Council of*

National Defense, 1917–1919 (Westport, Conn., 1984); Nancy R. Fritz, "The Montana Council of Defense," Master's thesis, Univ. of Montana, 1966; and Gerald Senn, "Molders of Thought, Directors of Action: The Arkansas Council of Defense, 1917–1918," *Arkansas Historical Quarterly* 36 (Autumn 1977): 280–90.

Arthur Link, "That Cobb Interview," *Journal of American History* 72 (June 1985): 7–17, discusses Wilson's feelings before April 1917 about what war might do to American freedoms. Harry N. Scheiber offers a critical account of what did happen in *The Wilson Administration and Civil Liberties, 1917–1921* (Ithaca, N.Y., 1960). Joan M. Jensen, *The Price of Vigilance* (Chicago, 1968), tells how the federal government employed the American Protective League to curb radicals and pacifists. Melvyn Dubofsky, *We Shall Be All: A History of the Industrial Workers of the World* (Chicago, 1969), presents detailed information about the Wilson administration's attacks on the Industrial Workers of the World. In *Politics is Adjourned: Woodrow Wilson and the War Congress 1916–1918* (Middletown, Conn., 1966), Seward W. Livermore shows what the president was up against as his Republican opponents attempted to use the war for their own purposes, a tendency from which Democrats were not exempt.

Those who want to understand why the war hysteria developed when and where it did should read John Higham's *Strangers in the Land: Patterns of American Nativism, 1860–1925* (New York, 1972), which explains the historical and psychological contexts of efforts to repress pacifists, radicals, and alleged German sympathizers. Stanley H. Coben, "A Study of Nativism: The American Red Scare of 1919–1920," *Political Science Quarterly* 79 (Mar. 1964): 52–75, uses an anthropological approach to explain both the rise of wartime hysteria and its metamorphosis into the postwar "red scare."

Chapter 3. The Managed Economy: Creating the Regulatory System

Kennedy, *Over Here*, lays out in a readable way the extraordinary complexities of economic mobilization. Weinstein, *The Corporate Ideal*, and Gabriel Kolko, *The Triumph of Conservatism: A Reinterpretation of American History, 1900–1916* (New York, 1977), examine developments in the American political economy that culminated in the mobilization system of 1917–18, while John P. Finnegan describes prewar foundations of that system in *Against the Specter of a Dragon: The Campaign for American Military Preparedness, 1914–1917* (Westport, Conn., 1975).

For cogent analyses of the way the leading war mobilization agencies operated and the reasons why they acted as they did, see Robert D. Cuff, *The War Industries Board: Business-Government Relations during World*

War I (Baltimore, 1973), and Paul A. C. Koistinen, "The 'Industrial-Military Complex' in Historical Perspective: World War I," *Business History Review* 41 (Winter 1967): 378–403. This seminal article is reprinted in Koistinen, *The Military-Industrial Complex: A Historical Perspective* (New York, 1980). Cuff and Koistinen correct myths about the operation of the War Industries Board presented in *American Industry in the War* (New York, 1941), by Bernard M. Baruch, himself a self-proclaimed myth destroyer. Baruch's story is told well by Jordan Schwarz, *The Speculator: Bernard Baruch in Washington, 1917–1965* (Chapel Hill, N.C., 1981). Daniel R. Beaver describes the War Department's responses to demands to rationalize military procurement in *Newton D. Baker and the American War Effort, 1917–1919* (Lincoln, Nebr., 1967).

Two invaluable sources on the running of the war economy are the War Industries Board minutes, published as U.S. Congress, Senate, Special Committee Investigating the Munitions Industry, *Munitions Industry: Minutes of the War Industries Board from August 1, 1917 to December 19, 1918*, 74th Cong., 1st sess., 1935, Senate Committee Print No. 4; and Grosvenor B. Clarkson's *Industrial America in the World War: The Strategy Behind the Lines, 1917–1918*, rev. ed. (Boston 1924), by a public relations man who served with the Council of National Defense. Clarkson's account is read most usefully between the lines.

Aaron A. Godfrey, *Government Operation of the Railroads* (Austin, Tex., 1974), and K. Austin Kerr, *American Railroad Politics, 1914–1920: Rates, Wages and Efficiency* (Pittsburgh, 1968) explain the advantages railroad companies gained from federally supervised self-regulation. Railroad administrator McAdoo's other activities, as secretary of the Treasury, and their results are analyzed in Charles Gilbert, *American Financing of World War I* (Westport, Conn., 1970). Gilbert finds serious problems with the ways in which the Wilson administration taxed and borrowed to pay for the war, for instance, the way wartime financing increased the purchasing power of certain lower-income groups while aggregate savings declined. In "Herbert Hoover, the Ideology of Voluntarism and War Organization During the Great War," *Journal of American History* 64 (Sept. 1977): 358–72, Robert Cuff explains why the Food Administration head preferred to employ volunteers; for instance, the way a volunteer system allowed elite leaders like Hoover the greatest freedom to run their organizations as they saw fit. James P. Johnson examines the origins and activities of the Fuel Administration in *The Politics of Soft Coal: The Bituminous Industry from World War I Through the New Deal* (Champaign, Ill., 1979), showing why the soft-coal industry was less successful than others in arranging to regulate itself.

For the government's management of labor, see Valerie Jean Conner,

The National War Labor Board: Stability, Social Justice, and the Voluntary State in World War I (Chapel Hill, N.C., 1983), Simeon Larson, *Labor and Foreign Policy: Gompers, the AFL, and the First World War* (Rutherford, N.J., 1974), and Harold M. Hyman's account of a federal "company union," *Soldiers and Spruce: Origins of the Loyal Legion of Loggers and Lumbermen* (Los Angeles, 1963). In *The Fall of the House of Labor: The Workplace, the State, and American Labor Activism, 1865–1925* (Cambridge, Mass., 1987), David Montgomery shows how some workers attempted to make lasting gains in wartime, defying Samuel Gompers's pledge not to change the status quo.

Chapter 4. The War Economy: Motivations and Results

Most of the sources for this chapter are the same as for the previous one. Melvin I. Urofsky, *Big Steel and the Wilson Administration: A Study in Business-Government Relations* (Columbus, Ohio, 1969), examines the administration's efforts to bring steel prices and distribution under its control. As its subtitle indicates, Urofsky's work analyzes the federal government's relationships with powerful corporate leaders before and during the war, a subject also treated in the works of James Weinstein and Gabriel Kolko mentioned earlier.

The War Industries Board's efforts to control automobile production are documented in the WIB minutes. André Kaspi, *Le Temps des américains: Le Concours américain à la France en 1917–1918* (Paris, 1976), sums up America's economic contributions to the Allies. On demobilization and the prospects for continuing the war welfare state into peacetime, see Benedict Crowell and Robert F. Wilson, *Demobilization: Our Industrial and Military Demobilization after the Armistice, 1918–1920* (New Haven, 1921), the March 1919 issue of *The Annals* (vol. 82), and especially Robert F. Himmelberg, "The War Industries Board and the Antitrust Question in November 1918," *Journal of American History* 52 (June 1965): 59–74.

Chapter 5. The War and Social Reform: Workers and the Poor

A comprehensive discussion of the effects of the war on social-welfare movements and organizations appears in Allen F. Davis, "Welfare Reform and World War I," *American Quarterly* 19 (Fall 1967): 516–33. However, one should also see John F. McClymer, *War and Welfare: Social Engineering in America, 1890–1925* (Westport, Conn., 1980), and Christopher Lasch, *The New Radicalism in America: The Intellectual as a Social Type* (New York, 1965), for a less rosy appraisal of what welfare reformers made out of the war. This chapter used several articles from the social workers' journal, *Survey*.

The works of Larson, Conner, and Montgomery listed above discuss relations between organized labor and the Wilson administration. For the complexities of workers' responses to the war, see also Melvyn Dubofsky, "Abortive Reform: The Wilson Administration and Organized Labor, 1913–20," and David Montgomery, "New Tendencies in Union Struggles and Strategies in Europe and the United States, 1916–1922," both in *Work, Community, and Power: The Experience of Labor in Europe and America, 1900–1925*, edited by James E. Cronin and Carmen Sirianni (Philadelphia, 1983). Maurine Weiner Greenwald discusses the varied effects of the war on female employees in *Women, War, and Work: The Impact of World War I on Women Workers in the United States* (Westport, Conn., 1980). Miles L. Colean, *Housing for Defense*. . . (New York, 1940), and Roy Lubove, "Homes and 'A Few Well Placed Fruit Trees': An Object Lesson in Federal Housing," *Social Research* 27 (Jan. 1961): 469–86, discuss the housing programs of the army, the Department of Labor, and the Emergency Fleet Corporation.

Chapter 6. The Great War and the Equality Issue: African-Americans and Women

A large literature is emerging on African-Americans in World War I. The attitudes of Wilson and his administration toward blacks in general and black troops in particular can be traced in Nancy J. Weiss, "The Negro and the New Freedom: Fighting Wilsonian Segregation," *Political Science Quarterly* 84 (Mar. 1969): 61–79, August Meier and Elliott Rudwick, "The Rise of Segregation in the Federal Bureaucracy," *Phylon* 28 (Summer 1967): 178–84, and especially in *The Papers of Woodrow Wilson*. W. E. B. Du Bois's fluctuating views toward the war can be traced in the pages of the *Crisis* and in Elliott M. Rudwick, *W. E. B. Du Bois: Voice of the Black Protest Movement* (Urbana, Ill., 1982). Two works mentioned earlier (Theodore Kornweibel's article and Gerald R. Gill's dissertation) examine disaffection toward the war among African-Americans. Gill also discusses the pressure the administration placed on black dissenters.

Arthur E. Barbeau and Florette Henri have written a thoroughly researched analysis of the activities of black servicemen in their aptly titled *The Unknown Soldiers: Black American Troops in World War I* (Philadelphia, 1974), a subject more recently discussed in Bernard Nalty, *Strength for the Fight: A History of Black Americans in the Military* (New York, 1986). Kaspi's *Le Temps des américains* relates the employment of black troops to the politics of Allied strategy, particularly the controversy over amalgamating American forces with those of Great Britain and France.

One of the best ways of discovering what happened to blacks who entered the American armed forces is through documents compiled by Nalty and Morris J. MacGregor in vol. 4 (*Segregation Entrenched, 1917–1940*) of their *Blacks in the United States Armed Forces: Basic Documents* (Wilmington, Del., 1977). I have also used the World War I Survey housed in the U.S. Army Military History Institute archives, Carlisle Barracks, Pa. (Part of this appears on microfilm, 39 reels, University Publications of America, Inc.) for evidence of black responses to military service and of the attitudes of white soldiers toward their black compatriots. Spelling has been standardized in quotations from World War I Survey documents, except in a few cases where spelling variations convey a special quality of the writing.

Little scholarship has been published on the impact of the First World War on the American suffrage movement. Much of the story can be pieced together from newspaper accounts and general histories, including Eleanor Flexner, *Century of Struggle: The Woman's Rights Movement in the United States* (Cambridge, Mass., 1975), the fifth volume of Elizabeth Cady Stanton, Susan B. Anthony, and Matilda J. Gage, eds., *History of Woman Suffrage* (1881–1922; reprint, Salem, N.H., 1985), and Carrie Chapman Catt and Nettie R. Shuler, *Woman Suffrage and Politics: The Inner Story of the Suffrage Movement* (1926; reprint, Seattle, 1969). Jacqueline van Voris has written a biography of the chief strategist of the National American Woman Suffrage Association—*Carrie Chapman Catt: A Public Life* (New York, 1987). The activities of militant suffragists are described by Inez Haynes Irwin, *The Story of Alice Paul and the National Woman's Party* (1921; reprint, Fairfax, Va., 1977), and Doris Stevens, *Jailed for Freedom* (New York, [1920]). Eric F. Goldman discusses suffragist tactics and their effects in "Progress by Moderation *and* Agitation," *New York Times Magazine*, June 18, 1961. Two additional sources for following the wartime suffrage campaigns are U.S. Congress, House, Committee on Woman Suffrage, *Extending the Right of Suffrage to Women: Hearings before the Committee on Woman Suffrage*, 65th Cong., 2d sess., January 3, 1918; and Judith Papachristou, *Women Together: A History in Documents of the Women's Movement in the United States* (New York, 1976).

Chapter 7. The Great War, Prohibition, and the Campaign for Social Purity

In *Deliver Us from Evil: An Interpretation of American Prohibition* (New York, 1976), Norman H. Clark explores the cultural roots of the American prohibition movement and links it with contemporary efforts to wipe out prostitution and with other efforts to use the state to cure social problems. James H. Timberlake, *Prohibition and the Progressive Movement, 1900–1920* (Cambridge, Mass., 1963), describes direct and indirect ways in which the war promoted restrictions on drinking. K. Austin Kerr documents the Anti-saloon League's increasingly successful tactics and propaganda in *Organized for Prohibition: A New History of the Anti-Saloon League* (New Haven, 1985). He shows, among other things, how Woodrow Wilson tried to avoid making a commitment to this pressure group.

David Pivar, *Purity Crusade: Sexual Morality and Social Control, 1868–1900* (Westport, Conn., 1973), analyzes the nineteenth-century origins of the wartime campaign to abolish prostitution. For an account by a leader of that campaign, see Raymond B. Fosdick, *Chronicle of a Generation: An Autobiography* (New York, 1958). There is a superb analysis of efforts to cope with prostitution during the war in Allan M. Brandt, *No Magic Bullet: A Social History of Venereal Disease in the United States since 1880* (New York, 1985), a subject that Fred D. Baldwin also examines in "The Invisible Armor," *American Quarterly* 16 (Fall 1964): 432–44, and at greater length in his 1964 Princeton Univ. diss., "The American Enlisted Man in World War I."

George Walker tells the official story of *Venereal Disease in the American Expeditionary Forces* (Baltimore, 1922). Frank Tannenbaum's lament about the morals of his comrades comes from the *Dial*, April 5, 1919. Two important documentary sources are U.S. War Department, *Home Reading Course for Citizen-Soldiers* (Washington, D.C., 1917)—for the kind of sexual codes the army wanted to inculcate in the troops; and the World War I Survey for indications of how they actually thought and behaved. I read all diaries, memoirs, letters home, and extended comments attached to questionnaires of the microfilmed version of the Survey produced by University Publications of America; the responses to the questionnaire by every fifth officer and noncommissioned officer and by every tenth enlisted man; and other materials in the World War I Survey and other Army Military History Institute collections that have not been microfilmed.

Chapter 8. American Intellectuals and the Control of War: Dewey, Lippmann, and Bourne

Two works provide, respectively, the cultural and intellectual context for this chapter and the one that follows: Christopher Lasch, *The New Radicalism in America: The Intellectual as a Social Type* (New York, 1965), and Morton G. White, *Social Thought in America: The Revolt Against Formalism* (Boston, 1957).

For connections between the career and the ideas of John Dewey, see George Dykhuizen, *The Life and Mind of John Dewey* (Carbondale, Ill., 1973), Neil Coughlan, *Young John Dewey: An Essay in American Intellectual History* (Chicago, 1975), David W. Marcell, *Progress and Pragmatism: James, Dewey, Beard, and the American Idea of Progress* (Westport, Conn., 1974), and Gary Bullert, *The Politics of John Dewey* (Buffalo, N.Y., 1983). Dewey's writings referred to here are mostly from *Essays on Politics and Education 1916–1917*, vol. 10 of John Dewey, *The Middle Works, 1899–1924*, edited by Jo Ann Boydston (Carbondale, Ill., 1976).

For Walter Lippmann, I have used the brilliant biography by Ronald Steel, *Walter Lippmann and the American Century* (Boston, 1980), and Lippmann's writings, including *A Preface to Politics* (1914; reprint, Ann Arbor, Mich., 1962), *Drift and Mastery: An Attempt to Diagnose the Current Unrest* (1914; reprint, Englewood Cliffs, N.J., 1961), and pieces in the *New Republic*. Charles Forcey analyzes the sociology of that journal and the ideas of its contributors in *The Crossroads of Liberalism: Croly, Weyl, Lippmann, and the Progressive Era 1900–1925* (New York, 1961), while John A. Thompson gathers together the thoughts of Lippmann and several of his liberal contemporaries in *Reformers and War: American Progressive Publicists and the First World War* (Cambridge, 1987). The quotation about supplying the Battalion of Death comes from Lippmann, "Notes for a Biography," *New Republic* 63 (July 16, 1930). Lawrence E. Gelfand discusses Lippmann's ultimately futile work planning for a better world in *The Inquiry: American Preparations for Peace, 1917–1919* (New Haven, Conn., 1963). For the postwar reactions of people like Lippmann and Dewey, see Stuart Rochester, *American Liberal Disillusionment in the Wake of World War I* (University Park, Penn., 1977).

Carl Resek has edited Randolph Bourne's wartime essays in *War and the Intellectuals: Essays by Randolph S. Bourne, 1915–1919* (New York, 1964), which includes an interpretation of Bourne's state of mind differing from the view in the poem excerpted here from John Dos Passos, *Nineteen Nineteen*, vol. 2 of *U.S.A.* (New York, 1937). Bruce Clayton's *Forgotten Prophet: The Life of Randolph Bourne* (Baton Rouge, La., 1984) is the standard biography which connects Bourne's thinking to the milieu in which he lived and wrote. While Clayton feels Bourne was not alienated, he includes evidence that suggests otherwise.

Chapter 9. The University at War: Veblen, Yerkes, Beard, and Cattell

The best book on the academic profession in World War I is Carol S. Gruber's insightful and extensively researched *Mars and Minerva: World War I and the Uses of the Higher Learning in America* (Baton Rouge, La., 1976). I have relied a good deal on Gruber's findings.

Two incisive biographies of Veblen that offer clues to his thinking about the Great War are John P. Diggins, *The Bard of Savagery: Thorstein Veblen and Modern Social Theory* (New York, 1978), and Joseph Dorfman, *Thorstein Veblen and His America* (New York, 1966). See also the superb brief account of this eccentric genius in Robert L. Heilbroner, *The Worldly Philosophers: The Lives, Times, and Ideas of the Great Economic Thinkers* (New York, 1953), and David B. Danbom's article, "'For the Period of the War': Thorstein Veblen, Wartime Exigency, and Social Change," *Mid-America* 62 (Apr.-July 1980): 91–104. The most pleasurable way of approaching Veblen is through his own writing, including *The Theory of the Leisure Class* (New York, 1934), *The Higher Learning in America; A Memorandum on the Conduct of Universities by Business Men* (1918; reprint, New York, 1965), and, especially for the purposes of this chapter, *Imperial Germany and the Industrial Revolution* (1939; reprint, New York, 1964).

A fine analysis of the varying aims and achievements of Yerkes, Cattell, and other psychologists during the period covered here can be found in Thomas M. Camfield, "Psychologists at War: The History of American Psychology and the First World War," Ph.D. diss., Univ. of Texas at Austin, 1969. Several articles examine the history of testing and the activities of Yerkes and Cattell just before and during the war, including Franz Samelson, "Putting Psychology on the Map: Ideology and Intelligence Testing," in *Psychology in Social Context*, edited by Allan R. Buss (New York, 1979), 103–68; and "World War I Intelligence Testing and the Development of Psychology," *Journal of the History of the Behavioral Sciences* 13 (July 1977): 274–82; Daniel Kevles, "Testing the Army's Intelligence: Psychologists and the Military in World War I," *Journal of American History* 55 (Dec. 1968): 565–81; and a series of papers by Michael M. Sokal, Richard T. von Mayrhauser, James Reed, and Henry Minton in *Psychological Testing and American Society 1890–1930*, edited by Michael M. Sokal (New Brunswick, N.J., 1987). Gruber discusses the events that led to Cattell's undoing in *Mars and Minerva*. For some of the ideas that enraged Columbia's president and trustees and a number of Cattell's colleagues, see James McKeen Cattell, *University Control* (1913; reprint, New York, 1977).

In *Mars and Minerva*, Gruber describes the behavior of historians in wartime America with understanding if not sympathy. George T. Blakey indicts them in *Historians on the Homefront: American Propagandists for the Great War* (Lexington, Ky., 1970). For long-term intellectual and cultural developments within the historical profession of which the events of 1917–18 were a crucial part, see Peter Novick, *That Noble Dream: The "Objectivity Question" and the American Historical Profession* (Cambridge, 1988). Samples of the historians' war work include Beard's article, "Atrocities," in Frederick L. Paxson, Edward S. Corwin, and Samuel B. Harding, eds., *War Cyclopedia: A Handbook for Ready Reference on the Great War* (Washington, 1918), and Wallace Notestein and Elmer Stoll's pamphlet *Conquest and Kultur,* mentioned earlier.

The discussion in this chapter of Charles Beard's career and personality derives from Richard Hofstadter, *The Progressive Historians: Turner, Beard, Parrington* (New York, 1968); from Ellen Nore's *Charles A. Beard: An Intellectual Biography* (Carbondale, Ill., 1983); and from a brief portrait, "Charles A. Beard: An Impression," by Eric F. Goldman, a former student of Beard's, in *Charles A. Beard: An Appraisal,* edited by Howard K. Beale (Lexington, Ky., 1954), 1–7. For accounts of Cattell's dismissal and the responses of Dewey and Beard, see Walter P. Metzger, *Academic Freedom in the Age of the University* (New York, 1964), Gruber, *Mars and Minerva*; Dykhuizen, *Dewey*; Nore, *Beard*; and Beard's letter of resignation from Columbia in *School and Society* 6 (Oct. 13, 1917): 446–47.

Chapter 10. The Battleground

The most astute analysis of the impact of the American army on the war appears in Kaspi's *Le Temps des américains.* An excellent general account of the raising of the American armed forces for World War I, their training, and the actions they took part in, with particular emphasis on the American Expeditionary Force, is Edward M. Coffman, *The War to End All Wars: The American Military Experience in World War I* (1968; reprint, Madison, Wis., 1986). Coffman notes the difficulty in finding accurate and agreed-upon military statistics for that war. Nevertheless, *The War with Germany: A Statistical Summary* (Washington, 1919) by Leonard P. Ayres, the chief of the War Department General Staff Statistics Branch, provides at least approximate numbers for training, supplies, weapons, costs, casualties, and other significant military matters. Additional data, brief accounts of battle actions, and superbly detailed maps appear in American Battle Monuments Commission, *American Armies and Battlefields in Europe: A History, Guide, and Reference Book* (Washington, 1938).

John P. Finnegan, *Against The Specter of a Dragon* (mentioned earlier), discusses prewar efforts to build up the American armed forces. John Whiteclay Chambers II has written the best account of conscription in World War I in *To Raise an Army: The Draft Comes to Modern America* (New York, 1987).

Two works about the American army in France that focus on the experiences of men in combat are Laurence Stallings, *The Doughboys: The Story of the AEF, 1917–1918* (New York, 1963)—by a participant; and Frank Freidel, *Over There: The Story of America's First Great Overseas Crusade*, rev. ed. (New York, 1990). Firsthand accounts of the battleground appear in the numerous biographies, diaries, personal narratives, and fictional accounts listed in the Schaffer and Woodward and Maddox bibliographies and in the New York Public Library's *Subject Catalog of the World War I Collection*, 4 vols. (Boston, 1961). Most information in this chapter about combat conditions and life behind the lines comes from the World War I Survey and from the published narratives claiming to be nonfiction.

Chapter 11. Motivating the AEF

William Langer's comment on the willingness of American troops to fight comes from his *Gas and Flame in World War I* (New York, 1965). Thomas Camfield has written about efforts to motivate the AEF in "'Will to Win'—The U.S. Army Troop Morale Program of World War I," *Military Affairs* 41 (Oct. 1977): 125–28. Other sources for this chapter are chiefly those for chapter 10, together with the poster collections noted for chapter 1, the War Department's *Home Reading Course*, Fred Baldwin's diss. noted in chapter 7, John P. Finnegan's *Against the Specter of a Dragon* on prewar training camps, and Chambers's *To Raise an Army* on conscription.

Information about the army's efforts to convert conscientious objectors can be found in Stephen M. Kohn, *Jailed for Peace: The History of American Draft Law Violators, 1658–1985* (Westport, Conn., 1986), David D. Lee, *Sergeant York: An American Hero* (Lexington, Ky., 1985), and Mark A. May, "The Psychological Examination of Conscientious Objectors," *American Journal of Psychology* 31 (Apr. 1920): 152–65.

Chapter 12. The Treatment of "Shell-shock" Cases in the AEF: A Microcosm of the War Welfare State

The chief source for this chapter is an official history, Pearce Bailey et al., eds., *Neuropsychiatry*, vol. 10 of *The Medical Department of the United States Army in the World War* (Washington, 1929), which contains numerous illustrative case studies, statistical reports on the incidence of psychological and neurological conditions in the AEF, Dr. Salmon's report on the treatment of mental illness and "war neurosis" in the British army, two postwar follow-up studies of treated shell-shock casualties, a large bibliography, and considerable information suggesting how army psychiatrists felt about their patients. Much of the material in *Neuropsychiatry* on the treatment of shell-shock casualties also appears in Edward A. Strecker, "Military Psychiatry: World War I, 1917–1918," in American Psychiatric Association, *One Hundred Years of American Psychiatry* (New York, 1944), 385–416. Abram Kardiner and Herbert Spiegel include several case studies from World War I in *War Stress and Neurotic Illness* (New York, 1947), and there are a number of descriptions of what soldiers called shell shock in the World War I Survey along with several reports of what would today be called post-traumatic stress disorder. The *Annual Report of the Administrator of Veterans Affairs, 1940* (Washington, 1941) and George K. Pratt, *Soldier to Civilian: Problems of Readjustment* (New York, [1944]), present statistical information about surviving victims of mental illness who served in the Great War.

Lawrence Ingraham and Frederick Manning describe the connections between American civilian and military psychiatry and survey the treatment of psychological casualties, including those from the First World War, in "American Military Psychiatry," *Military Psychiatry: A Comparative Perspective*, edited by Richard A. Gabriel (New York, 1986), 25–36. Gerald Grob examines the mental-hygiene movement which influenced the treatment of World War I shell-shock victims in *Mental Illness and American Society, 1875–1940* (Princeton, N.J., 1983). For additional information about American psychiatry in the World War I era, see John C. Burnham, *Psychoanalysis and American Medicine: 1894–1918; Medicine, Science, and Culture* (New York, 1967), and the journal *Mental Hygiene*.

Epilogue

William E. Leuchtenburg discusses various ways in which Franklin Roosevelt and his administration applied wartime methods to Depression problems in "The New Deal and the Analogue of War," in *Change and Continuity in Twentieth-Century America*, edited by John Braeman, Robert H. Bremner, and Everett Walters (Columbus, Ohio, 1964), 81–143.

Index

African-American military personnel, 75–76, 80–89; combat performance, 86–89; discrimination against in France, 85–87; history of, ignored, 89; morale, 84; and racial attitudes of white army officers, 82–83, 84–87

African-Americans: attitudes toward war, 77; cooperation in war effort, 80; gains and losses from war, 89–90; migration from rural south, 75; protest discrimination and racial violence, 77; and racial violence, 75–76

American Alliance for Labor and Democracy (AALD): as CPI front organization, 5–6; fights People's Council and IWW, 67; Gompers as nominal head of, 6; publishes propaganda, 7

American Association of University Professors (AAUP), and wartime limits on academic freedom, 146

American Expeditionary Force (AEF): campaigns, 150–51; makeup of, 150; and race relations, 85; and social purification, 104–5

American Federation of Labor (AFL): cooperates with Wilson administration, 66–67; encourages labor support for administration foreign policy, 7, 67; fights pacifists and radical labor organizations, 66–67; strikes bargain with Wilson administration, 67–68

American Protective League: cooperates with Department of Justice, 17, 27–28; operations against dissenters, 17

Anti-suffragists, 91, 94

Antitrust laws: Gregory wants to enforce, 48–49, 59–60; and Railroad Administration, 37; and War Industries Board, 45; Wilson accepts suspension of, 60

Baker, Newton D.: and African-Americans, 81, 81n; decries lack of public spirit by labor and management, 49; and "invisible armor," 100; and Lippmann, 115; organizes popular support for conscription, 176; as reformer,

24; as social purity advocate, 100, 104; and War Department hoarding, 32

Ballou, Charles C., attitude toward African-American troops, 83–84, 86–88

Bargaining: between AFL and Wilson administration, 66–68; between African-Americans and Wilson administration, 77, 80; between automobile industry and War Industries Board, 52–53, between National American Woman Suffrage Association and U.S. government, 92–93, 94; between shell-shock casualties and army medical personnel, 206; between social reformers and government, 108; between steel industry and War Industries Board, 56–57

Baruch, Bernard M.: appointed head of War Industries Board, 43; character and behavior, xiv, 43–44; exploits limited authority, 44–45; fears public will see through War Industries Board myths, 63; indirect approach of, 43; mastery of public relations, 44; and myth of cooperation by commodity producers, 50

Battle: avoidance of, 166–67; effects on troops, 172, 174; horrors, 156–59; reluctance of soldiers to talk about, 173–74; sights, 156–57, 164, 167; smells, 158; sounds, 157–58

Beard, Charles A., 118; advocates U.S. participation in world war, 139, 141–42; character and career, 141; dislikes idealism of American propaganda, 142–43; resigns from Columbia University, 147–48; sides with Allies, 142; uses history to reform society, 141–42; war work of, 142

Bettman, Alfred: ambivalence toward repression, 27–28; fails to curb federal district attorneys, 28; sanctions American Protective League activities, 28

Bigelow, Herbert S., abducted by vigilantes, 24

Bourne, Randolph: antiwar outlook of, explained, 123–24; attacks pro-war intellectuals, 120–22; biography and character, 118; critic of middle-class culture,

237